T5-ASN-586

BEYOND TAYLORISM

Beyond Taylorism

Computerization and the New Industrial Relations

Lorraine Giordano

Project Director
National Center for Research in Vocational Education
University of California, Berkeley

St. Martin's Press

331.25
G49b
c.2

© Lorraine Giordano 1992

All rights reserved. No reproduction, copy or transmission of
this publication may be made without written permission.

No paragraph of this publication may be reproduced, copied or
transmitted save with written permission or in accordance with
the provisions of the Copyright, Designs and Patents Act 1988,
or under the terms of any licence permitting limited copying
issued by the Copyright Licensing Agency, 90 Tottenham Court
Road, London W1P 9HE.

Any person who does any unauthorised act in relation to this
publication may be liable to criminal prosecution and civil
claims for damages.

First published in Great Britain 1992 by
THE MACMILLAN PRESS LTD
Houndmills, Basingstoke, Hampshire RG21 2XS
and London
Companies and representatives
throughout the world

A catalogue record for this book is available
from the British Library

ISBN 0–333–55113–3

Printed in Great Britain by
Antony Rowe Ltd, Chippenham, Wiltshire

First published in the United States of America 1992 by
Scholarly and Reference Division,
ST. MARTIN'S PRESS, INC.,
175 Fifth Avenue,
New York, N.Y. 10010

ISBN 0–312–07579–0

Library of Congress Cataloging-in-Publication Data
Giordano, Lorraine, 1947–
Beyond Taylorism: computerization and the new industrial
relations / Lorraine Giordano
    p.   cm.
Includes bibliographical references and index.
ISBN 0–312–07579–0
1. Labor supply—Effect of automation on.   2. Manufacturing
processes—Automation.   3. Manufacturing processes—Data processing.
4. Industrial relations.   5. Quality circles.   6. Industrial
management—Employee participation.   I. Title.
HD6331.G56   1992   91–38759
331.25–dc20   CIP

Contents

University Libraries
Carnegie Mellon University
Pittsburgh PA 15213-3890

This book is dedicated to all the IUE members who so generously shared their experiences and ideas and without whom this research could not have been completed.

la lotta continua

Acknowledgements

This book is based on research for my doctoral dissertation. The arduous task of completing the degree was made easier by the advice and support of colleagues and friends in academia and the labour movement.

The IUE generously provided access to members whose firm met the specifications required for the research. Bob Kennedy, the union's international representative, established contacts for me with each of the locals' leadership giving the project his full support and saving it from sure disaster when one of the presidents changed his mind about his local participating in the study. I owe him a very special debt of gratitude.

As with all research, this project went through numerous stages of formulation. My advisor, Wolf Heydebrand, encouraged my explorations and challenged me to clearly develop my ideas. When I concentrated on the 'larger issues' Patricia Sexton pointed out the concrete details of doing research which had to be carefully planned if the study was to be successful. Her counsel always proved correct. Ruth Milkman enthusiastically supported the research from the outset. Her interest and invaluable comments have contributed enormously to this work. Beth Stevens and Edward Lehman provided cogent recommendations for improving the clarity of the manuscript. Wanda Orlowski's ideas and work on information technology were extremely insightful and helpful.

My dissertation group was ultimately responsible for seeing me through the completion of this work. Sophia Catsambis' incredible organizational abilities and her simple explanations consistently solved problems which appeared insurmountable to me. Linda Cushman's keen insights and clear judgment always provided needed perspective when issues appeared to be too complex. Cherni Gillman, with her capable editing skills and quick wit, formulated and presented ideas in a new light. Feminist praxis was always in evidence as we interwove discussions of our work, the problems we encountered, and personal triumphs we shared and celebrated.

Nancy Addess provided very capable editorial work and suggestions. T. M. Farmiloe, my editor, made the process of publishing this book almost effortless. My colleagues at Queens College, sensitive to the difficulties facing adjuncts, provided support and interest in my work as well as steady employment. Over my protests, they encouraged me to conquer my own fears of computers and gave me unlimited access to their system. Judy Fawcett's kindness and very able transcription skills are much appreciated and I am grateful for her generosity. In characteristic fashion all the members of my family provided encouragement throughout this process. With humour, concern and pride, they gave me their full support.

Finally, my friends deserve a special note of thanks. Roberta Franck was there with much needed understanding and diversions. Cathie Doyle gave me wise counsel and unconditional friendship. Miriam Frank showed me that a sense of humour was vital for surviving a dissertation. Barbara Beno provided the physical and emotional space to complete this book. I am grateful for their friendship.

1 Introduction

The last decades of the twentieth century have seen rapid changes in the way work is performed, where it is done, the kinds of work available to people, and labour-management relations. Industrial organization is being re-shaped by two distinct yet inter-related processes – computerization and new labour relations policies, among them labour-management participation programmes. Popularly known as Quality of Work Life, or QWL, these programmes are organizational strategies designed to address production, motivation and control on the shopfloor and usually take the form of quality circles (QCs) in which small groups of workers and a supervisor discuss issues of quality and productivity.[1]

Automation and QWL represent current industrial attempts to reorganize production methods. In the broadest – and most crucial – sense, they are part of management's efforts to secure profits by reducing labour costs and increasing predictability in production. Historically, management has always adopted technical and social methods which would maintain, and preferably increase, control over production. As Nelson points out, the factory system that emerged in the US between 1880 and 1920 was both the product of mass production techniques and the result of changes in management practices. Social welfare programs and the development of a systematic supervisory system were an integral part of this development.[2]

In much the same way, contemporary strategies for collective bargaining and negotiated settlements, automation, and decentralization reflect a shift toward an increasingly global and interdependent economy. Intensified competition within this more tightly integrated economy, shifts in labour markets, and the demand for specialized products contribute to corporate restructuring and the reorganization of the labour process.

Automation and Quality Circles are examined here in terms of current industrial and managerial changes and as inter-related processes. The depth of these changes depends on the extent to which: (1) industry can utilize fully automated systems in its production processes; (2) automation is used to reorganize existing production methods; (3) the industry is decentralizing operations and/or centralizing aspects of its planning and decision-making; (4) top management is able to engage the cooperation of middle and lower-level managers who may feel their jobs are in jeopardy; (5) upper level management will, in turn, support decisions

1

made on the local level; (6) management depends upon and can gain
workers' cooperation in reorganizing production; and (7) workers resist
these efforts, the strength of their organization, and past history of labour
relations.

This book explores the concrete ways automation and Quality Circles
affect the labour process through a detailed case study of a defence
plant. The questions of skill, control and hierarchy inform this work.
However, the emphasis of the analyses of these processes is on the
relational aspects of these processes rather than on traditional categories
of efficiency, rationality and integration.

The corporation analyzed in this work is a major producer of defence-
related equipment in the US. Originally a producer of specially designed
instruments for military, industrial and farming equipment, it now concentrates more of its operations on the design, engineering and development
of computer-related products for the US Department of Defense and select
commercial markets. It has major facilities throughout the United States,
and in Canada, Europe and South America. For the purposes of this study,
the corporation will be referred to as UFC.[3] The plant studied is a low,
sprawling complex, surrounded by pine trees and shrubs, located just
outside of a major north-eastern city off a major highway. I will call
the facility Pine Hill.

The workers interviewed are members of three different locals of the
International Union of Electrical, Radio, Machine and Technical Workers
Union (IUE) in the plant. In this book, the local representing machinists
will be referred to as Local 3; the local representing process engineers as
Local 8; and the drafters' and design drafters' local as Local 5.

This plant was chosen for several reasons. (1) The equipment used
in the plant ranges from old conventional machine tools to numerical
control (NC) and state-of-the-art computerized numerical (CNC) machine
tools. (2) Drafting and design work continues to be done manually at the
drafting board as well as on computer-aided design (CAD) equipment. The
drafting and design department has worked with several 'generations' of
CAD equipment. Workers' ten years of experience with computerized
equipment gives them an opportunity to assess any changes which have
taken place in their work. UFC's Pine Hill facility was the first firm in
the area to purchase CAD and has the most up-to-date system, the CADD
4X. (3) The QC program had been in effect for over a year at the time
the study began, long enough for any 'Hawthorne effect' to be reduced
or eliminated, making it possible to assess its failures and successes. (4)
It is a union shop. Any changes management wants to introduce in the
organization of work have to be explained to and negotiated with the

union. This provides an opportunity to analyze the impact of changes on labour relations and on the shopfloor.

This study focuses on four of the occupations at the plant – machinists, process engineers, drafters and design drafters. These occupations are related in the production process and each is responsible for a different phase in design and production work. Designers develop a concrete and accurate image of an engineering concept. Drafters are responsible for presenting this work in an accurate, blueprint format. Process engineers determine the machining instructions for part production and assembly which conform to the drafters' specifications, and machinists, using these instructions or working directly from the blueprint, produce the finished part.

Since computerization is primarily an information system, its effects on occupations and the labour process go beyond the mechanization of manual work. Accessibility to a range of information creates the potential for shifts in the demarcation between traditionally defined occupations.

In previous analyses of skill changes brought about by automation, Braverman[4] has emphasized the de-skilling thesis and Blauner developed the notion of the 'reliable employee' based on his comparative study of technology, skill and autonomy.[5] However, the processes of de-skilling and the introduction of new skills and/or occupations are *concurrent processes*. At any given moment, the labour process embodies de-skilling, skill development, conflict and cooperation. Thompson identifies these tendencies within production as 'unfinished processes' which are historically determined.[6] A comprehensive analysis of skills and the organization of work would take into account existing technical and social divisions of labour, the history of labour relations within a firm, market conditions, and the competitive strength of the firm.[7]

Within this framework, it is important to consider the fact that computerization affects both process *and* product, significantly impacting on what is to be produced, rapidly changing market needs, as well as the methods used to produce those goods. Technology, therefore, represents one aspect in a multi-dimensional analysis of skills and the organization of work.[8]

Strategies for managing changes in work redesign include gaining the cooperation of the workforce. QC is part of this reorganization process, both changing the organization of work and soliciting ideas for increasing or maintaining a competitive edge within the market. Quality Circles are, in actuality, part of the 'contested terrain' of workplace control. As such, its reliance on workers' cooperation suggests a capitulation by management to workers' demands for increased participation as well as a recognition of their knowledge of the production process and the informal organization on the shopfloor. The use of informal work groups both to solve production problems and to act as a mechanism for discipline and cooperation suggests workers' resistance to both the process and outcomes. Although Quality Circles remain embedded within the traditional bureaucratic hierarchy, they also operate, to a degree, outside of it in the context of an informal decision-making body. The existence of such an informal mechanism for decision-making contradicts Edwards' thesis of separate and successive forms of control.[9] Instead, it supports Thompson's notion of a 'combination of managerial structures'.[10]

QWL is a technique that not only encourages cooperation (although it is a critical aspect of flexible manufacturing), but also enhances coordination of complex tasks within these new production methods. It is first and foremost an *organization* tool designed to integrate operations which will increasingly depend on skill, judgment, and initiative. The widespread application of computers and computerized equipment, including robotics, to all phases of production, office automation, and communication systems integrates skills, occupations, and information demanding collective strategies in order to achieve production goals.

QWL is designed to facilitate consensus. However, the democratic tendencies within this process are mitigated by management's right to define the agenda. Thus QWL ideology and its format of democratic, voluntary participation encourages both conformity to the process as well as a framework for challenging managerial prerogatives.

Workers' support for QWL appears to rest on the secondary outcomes of participation: (1) improved communication between workers and managers; (2) feelings of influence and control over aspects of the work process; and (3) actual influence on select production issues. To accept these outcomes as a primary and legitimate means toward achieving workplace democracy obscures the fundamental role of QWL as a decentralized *management* decision-making process which addresses the issues of cost, efficiency, productivity and product quality at the point of production. Moreover, the terms of worker participation are circumscribed largely by management's production needs, that is, improving productivity and efficiency. These are *primary* QWL objectives. Job satisfaction and a general improvement in the workers' quality of life are also important aspects of QWL. However, they are not pursued simply as ends in themselves, but as contributing to overall improvements in productivity.

Quality Circles and automation can, therefore, be examined as forms of both organization and control of the labour process.

MICROPROCESSORS AND THE LABOUR PROCESS

Two features distinguish computer automation from its predecessors: (1) the flexibility of its application to all phases of the production process; and (2) the capacity to store and generate the different types of information gathered in that process. Earlier forms of mechanization and mass-production techniques introduced in the 1920s replaced some of the physical labour and increased the speed and improved the continuity of production. However, mass production also included the use of specialized tools and fragmented and extended the division of labour. Manual work, therefore, always remained a central element in production.[11] In the 1950s, cybernetics and continuous process production introduced high-speed processing to industries manufacturing liquid and gas products, in particular petroleum, chemical plants and food processing. Manual and skilled labour were then replaced with technical processes and industry-specific skills. The nature of the technology at that time limited its application almost exclusively to these industries.

Like previous technological innovations, microcomputer-based automation has eliminated jobs as well as created new categories of work. However, this form of automation differs from past technologies in its unique ability to link each phase of production through the computer system. It is both multi-functional and reprogrammable and can easily be adapted to changes in production needs.[12]

Computerization of both production and management information systems has several major consequences. First, it offers the means to eliminate whole series of operations by programming them into the computer, using robots, or recombining operations. Skills that are programmed are no longer the province of one particular occupation. These automated skills instead become a part of the general body of knowledge available throughout the internal apparatus of an industry, and can undermine workers' specific control over production processes – thus laying the groundwork for possible job elimination.

Second, a countervailing, but related, tendency is a shift in the lines of demarcation between jobs. Automated equipment's memory capabilities have been allowing workers to perform a wider range of operations. Computers can also simplify particular job functions which then makes it easier for workers in related occupations to perform them. However, acquisition of computer skills does not always translate into job upgrading or pay increases. While programming may represent additional skills, it may not be seen as comparable to those replaced by automation.

A third and related consequence of computerization is the possibility that some occupations will experience a radical downward shift in skills or qualifications necessary to perform them. A fourth consequence is that computerization offers greater flexibility in management control over the decision-making and production processes. According to Sanders, computerization has allowed management to use the same information for different purposes. At the executive level, it furnishes data for planning purposes, while middle management uses it primarily as a mechanism for control.[13]

The use of computerization in the decision-making process has contributed to the development of an industrial organization where planning and financial decisions are centralized and other operations are frequently decentralized and highly interdependent. Long-range planning for production and the coordination of interdependent operations are, increasingly, being centralized. However, with computerized production schedules, inventories, and product sales, decisions regarding on-site production issues, including product quality and the use of skills, are being delegated to local management. Decentralization also includes design of a structure that will aid corporate planning strategies and insure maximum compliance with scheduling and the coordination of production.

The connection between computerization and mechanisms for decision-making and control have been most directly linked by Herbert Simon and Peter Blau.[14] Focusing on the implications of automation for organizational structure, Simon examines the changes in authority relations within the managerial hierarchy. He argues that it is not a question of centralization *or* decentralization, but the forms each will assume.

Blau's work analyzes the effects of automation on the administrative structure. He concludes that the locus of decision-making varies according to the location of the computer. Organizations whose automated support functions are located in-plant experience greater autonomy at the local level, while those with computerized administrative functions at corporate headquarters have more centralized decision-making operations. Blau suggests that the availability of computers to perform administrative functions on the local level encourages the formation of multi-level hierarchies.

This kind of industrial reorganization requires the creation of new mechanisms for communication between corporate headquarters and its divisions. Among its goals are greater coordination of interdependent operations and the commitment of workers and managers to participate in this reorganization through their involvement in the set up and maintenance of such a mechanism. QWLs can be analyzed as a decentralized decision-making process which addresses the issues of cost, efficiency and product quality at the point of production. The contribution of workers' knowledge and skill to this process can be understood as part of the organization's information gathering process and performance predictability.

LABOUR-MANAGEMENT COOPERATION

The QWL movement has been associated with two well-publicized, yet very different issues: 'Japanese style' management and workplace democracy. American industrial folklore credits Japanese business with a visionary approach which has redirected adversarial labour-management relations toward a more cooperative accomplishment of goals.[15]

Workplace democracy is often cited as the only solution to job dissatisfaction and the 'right to dignity' on the job. What workers want, we are told, is more responsibility, more input and decision-making, more respect and less authoritarian control.[16] As abstract qualities, these are, in themselves, admirable and, according to democratic ideals, desirable. Few would argue with the need for greater participation in and control over the work in which we are engaged. Yet the use of the term 'workplace democracy' (or its equivalents, 'participative management' or 'work humanization') is misleading. Workplace democracy is, as Vaughan sees it, a 'catchword' implying consensus.[17] Assumed to be a shared experience, workplace democracy requires neither justification nor definition. Witte,[18] moreover, questions the validity of the use of the term 'democracy' as it applies to work under capitalism. Accepted notions of meritocracy and corporate rights to profits, along with varying degrees of influence and rewards within the production process, undermine the very principle of democratic organization.

There has been no lack of organizations promoting labour-management programmes. The US Department of Labor[19] lists over 200 centres, programmes and private organizations promoting such programmes, including university programmes, regional area programmes within the public sector, and joint labour-management programmes in the private sector. The federal government, under the Department of Labor and the Labor-Management Services Administration, created the Division of Cooperative Labor-Management Programs in 1982 to encourage and assist employers and unions to undertake joint efforts to improve productivity and enhance the quality of working life.[20] Previously, this work was carried out by the National Center for Productivity and the Quality of Work Life (formerly the National Commission on Productivity and Work Quality and under the Nixon-Ford Administrations, the Productivity Commission).[21]

Quality of Work Life (QWL) is a generic term representing a wide range of labour-management programmes. Specific types of programmes have been developed under such titles as Employee Involvement, Quality Circles, Job Enlargement, Job Enrichment, Work Humanization, Participative Management, Group Problem Solving and Autonomous Work Groups. Each type of programme is designed to encourage particular forms of work organization and decision-making, depending upon the nature of the industry and technology. For example, job redesign is prevalent in service sector industries, while autonomous work groups are often instituted in highly automated, continuous process plants.[22] However, as Wrenn notes, most writings on the subject do not make distinctions between the different programmes, and there is little, if any, clear definition of the terms and their meanings.[23]

QWL, however, was not 'made in Japan', nor is it a contemporary phenomenon born out of the frustrations of a dissatisfied and youthful workforce, as some proponents claim.[24] History provides evidence of similar attempts at gaining workers' cooperation, inspiring company loyalty, and avoiding work stoppages. Such past industrial inducements include the formation of the Whitley Councils and Joint Production Committees in Britain during both World War I and World War II, respectively; the Labor-Management Committees under the War Production Board in the United States during both world wars; and various profit-sharing plans and human relations projects and experiments during the early 1900s.[25]

Like earlier labour-management cooperation programmes, QWL emphasizes productivity, motivation and cooperation. What distinguishes these contemporary programmes from their historical counterparts are (1) the integration of Quality Circles into the reorganization process as corporations centralize control over planning and decentralize authority over local operations; and (2) the emphasis on the development of skills which are consistent with the needs of automated work processes – flexibility, independent judgment, initiative, and team work.

AUTOMATION, QUALITY CIRCLES AND THE WORK PROCESS

The technical and social divisions of labour, authority relations, and the specific uses of skills, knowledge and technology are all part of the social relations of production. Consequently, changes in any aspect of production will, by the nature of its relationship to the others, affect changes in all these areas. The forms and extent of these changes will depend on a number of factors, including the nature of the industry, worker resistance to these changes, production techniques, and the strength of the industry in the marketplace.

Historically, the struggle between labour and management over the right to determine and control the use of skills and the organization of work has shaped social relations. In this age of global capitalism, recentralization of corporate planning and increased autonomy of local management is structuring the production process and represents a contemporary strategy to assert managerial control.

Although automation and QC each affect different aspects of the labour process, they are, in essence, *two sides of the same process* and, therefore, inextricably linked. Within this context, automation and QC represent changes in the technical and social divisions of labour and, as such, are embedded in the existing social relations of an organization.

QC and automation significantly contribute to recentralization of planning and decentralization of authority over production decisions insofar as both are information-gathering and transmitting processes. The capacity for automated equipment to be used as an information system depends, to a great extent, upon workers' abilities and willingness to generate and fully utilize available computerized data. QC is a formal and systematic yet flexible means of accomplishing this goal. Data generated through computerized management information systems – known as MIS – provide local management with information on specific problems and goals concerning cost, scheduling and efficiency. The role of QC is to use this computer-generated information as it relates to efficient use of skills, product quality and the coordination of automated processes within the plant.

Because automation simplifies segments of the work process, management has the opportunity to redesign production according to its short and long-term needs. Job reclassification and the development of work teams promote a wider and more flexible application of skills in the use of computerized equipment. Through the problem-solving format of QC, management can promote and sustain changes in work organization. It encourages workers to draw on their experience and understanding of the production process and to be flexible in the use of their skills to create innovative solutions for cost, quality, and efficiency problems at the point of production.

The different production needs of industries determine both the kind and extent of automation installed in plants. There are two basic uses of computers in production. They can be applied to a range of operations to increase accuracy and versatility in the use of materials and methods without a major restructuring of the production process. Or they can be part of a larger automated system which, as Shaiken points out, directs production activities and 'will define, collect and control the processing of information throughout the factory'.[26] Both uses will increase the predictability of performance. However, fully automated industries tie together all of the production processes, often reorganizing work in very dramatic ways.

On the local level, computers provide managers with detailed data and the ability to spot potential problems in production flow and defects in parts or assembly. On the corporate level, these data are analyzed in relation to larger corporate goals and used to develop and assess strategies for increasing productivity, creating new products and coordinating production and delivery schedules.[27]

Dramatic changes in world trade patterns and the uses of computerization in both production and communications have forced corporations to adopt new mechanisms for responding to global market conditions. Kanter notes that in order to be successful, organizations 'will need to be able to bring particular resources together quickly, on the basis of short term recognition of new requirements and the necessary capacities to deal with them'.[28] Decisions on planning and resource allocation are increasingly being made on the corporate level which, Kanter indicates, 'shift the ways functions and units line up with respect to each other'.[29] Parallel structures have emerged which transcend traditional boundaries and create new patterns of managerial communication.

The implications of decentralization have extended to the executive suite with substantial reductions in corporate staff. According to Peters, these reductions are part of the second wave of corporate decentralization. The first occurred after World War II when industries expanded into different product lines and created separate divisions, each responsible for their own operations and markets. The second wave of 'radical' decentralization reduced the large centralized reporting system built up to oversee and monitor each division.[30] The shift toward local responsibility for specific production goals is part of this large movement to restructure the corporation. Decentralization allows decision making flexibility which complements and informs the corporation's long-range financial and planning strategies and will respond to unanticipated market changes or crises. Traditional departmental lines of communication and decision-making are being abandoned and regrouped, often according to product areas. Known as 'vertical linking', workers and management have two major responsibilities: to identify the problems and strengths of their operations, and to plan and implement the necessary changes in order to meet corporate goals.

Most noteworthy here is the growing importance of and reliance on line management to set up and maintain this system.[31] Their task is to create a mechanism for problem-solving and the motivation of their subordinates to participate in such a process. QC is often the vehicle used to carry out this process. Because it involves delegation of responsibility to subordinates and the blurring of conventional lines of authority, management has not always welcomed the change. Several factors contribute to managerial reluctance.

First, it pinpoints problems existing under a manager's or supervisor's domain. In traditional bureaucratic culture, this is viewed as a sign of ineffective leadership. Lower level management may feel vulnerable and may resist exposing those areas under their control which could be criticized as unproductive.

Second, this process requires workers to participate actively in the identification of problems as well as to initiate suggestions for improvement and change, further contributing to supervisors' feelings of expendability.[32]

Third, despite organizational commitment to the QC programme, upper level management may be uneasy with changes in long-standing corporate policies and management practices, resulting in little actual support for concrete proposals.

Ultimately, however, it is the anticipation of increased productivity and flexibility of the workforce that motivates management to establish new mechanisms of organizational decision-making and control. The countless uses of computerized information gathered from every facet of a company's operations can only be realized in an environment equipped to deal with them. Management knows that operators can either perform the basic functions to get a job done or, by working the information stored in the computer, find faster, less costly and improved production methods. No matter how sophisticated a computer system might be, its value as an information and production system remains limited if workers choose to restrict themselves to a minimal use of the data.

Certainly, this is nothing new in labour-management relations. Managers have always known that worker cooperation is essential, regardless of production methods. However, as Drucker states, there is a 'change from organizing production around the doing of things to things to organizing production around the flow of things and information'.[33]

Current research emphasizes either the impact of technology or the effects of neo-human relations programmes such as Quality Circles on organizational change (see Figure 1). However, *both* processes are part of a thrust for greater productivity and efficiency. While automation and Quality Circles are not necessarily adopted together, or directly linked together, both focus on the flexible use of skills and the transmission of information.

While the first two models provide insights into changes in the organiza-tion of work and in subsequent organizational behaviour, they suggest that each process is distinctly different and independent of the other. Moreover, control over the production process, which is part of the structure of social relations, is treated as an intervening variable, or a direct *effect* of technology.

The third model provides a more comprehensive view of the underlying causes of technological and managerial change. Rather than *beginning* with managerially instituted changes, this model suggests that existing social relations (at the point of production, at the level of the firm, and in wider economic conditions) influence the forms that technology and the organization of production and decision-making will assume. Both address the issues of predictability, efficiency, productivity and cooperation. How-ever, in this model, choices of technology and managerial programs such as QCs are the *outcomes* of organizational processes and need to be analyzed as such.

Current research analyzes the impact of computerization on occupations in terms of the elimination of traditional physical activities which have defined skilled crafts. The research presented here indicates, however, that in many instances some of these skills – and the processes associated with them – are transformed into conceptual work. For example, machinists no longer physically create a part using hand-practiced skills. Those conventional skills become abstract and conceptual processes. What is significant here is that *the method of applying skills is altered, not the skills themselves.*

Within this context of organizational change, specific issues surrounding the automation of work are examined here in terms of the following questions: To what extent does computerized design and production technology de-skill and/or re-skill jobs? What happens to the content of jobs as processes are automated and/or transformed into abstract applications of knowledge? Since this work becomes dependent upon computer-mediated information, how does the technology affect workers' ability to continue to exercise control over the use of their skills? How does management use the flexibility inherent in computerized production methods to reorganize the labour process? In particular, what does this flexibility mean for traditional job classifications and the type of work that becomes available when jobs are automated?

QC is studied here as part of a labour relations system that by definition involves conflict as a critical part of the relationship between labour

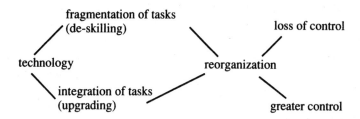

Model 1

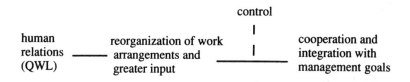

Model 2

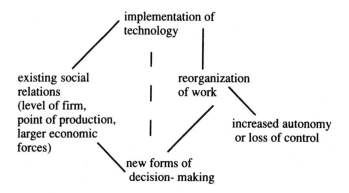

Model 3

FIGURE 1   *Changes in the organization of work*

and management. This work addresses the following questions: How do workers view the notion of 'cooperation'? Given the way QC is structured, what forms of participation emerge from this process? What issues are raised in QC sessions and how do they reflect the interests of labour and of management?

In this work, QCs are situated within the context of the new industrial labour relations. Contemporary collective bargaining strategies which are part of this process have focused on efforts to gain concessions from unions around such issues as the elimination of job classifications and categories, job rotation and concessions on wages and benefits. The emphasis of QC programs is on 'common goals' and 'cooperative relations'. However, these goals centre around the development of strategies for improving productivity and efficiency. This study examines the ways workers who participate in QCs understand these contradictions, resist efforts to comply with management's agenda and develop strategies of their own.

The book is divided into two main sections: (I) Automation and the Organization of the Labor Process, and (II) The New Industrial Relations and QWL. Chapter 2 is the description of the site and discussion of the methodology. Part I – Chapters 3, 4, 5, 6, 7 and 8 – describes, analyzes and compares workers' experiences with conventional and computerized equipment. Chapter 3 is a discussion of the literature on automation, skills and the organization of the labour process. There are essentially three positions within this debate: (1) technology as the primary contributor to either the de-skilling or re-skilling of occupations; (2) managerial control as the determinant of technological use in production; and (3) the influence of market and organizational forces such as the nature of industries, prevailing divisions of labour and labour relations on the introduction of automation and the resulting changes in skills.

Chapters 4, 5 and 6 examine the impact of automation on production workers at the plant – the machinists and process engineers. Chapter 4 discusses machinists'experiences working with three different types of machine tool technology: conventional, NC, and advanced CNC equipment. The traditional skills which have historically defined the craft are transformed as work is automated. Although still having a critical role in NC work, machinists have lost much of the direct control they exerted in conventional work. However, in programming the Hurco (CNC), machinists reclaim control over the process. With advanced computerized equipment, manual skills, experience and knowledge of machining principles are applied in an abstract format.

Chapter 5 looks at one of the industrial relations policies instituted by

UFC that is part of the reshaping of the workplace. With the automation of the machine shop and the redefinition of the Hurco as a conventional tool, management proposed the merger of job classifications in the shop. With the elimination of highly specialized manual skills, management gains increased flexibility in the use of its labour force, and minimizes both workers' control over the use of their skills and the power of the shop steward to negotiate over the use of these skills at the point of production.

Job classifications have historically protected workers against the fragmentation and intensification of the labour process. The elimination and simplification of some tasks and the reintegration of other processes increases the possibility for restructuring the organization of work, removing these protections. Along with job rotation and the elimination of work rules, the merger of job classifications can be viewed as part of an overall strategy to eliminate collective bargaining agreements (and with it, unions) and provide management with the opportunity to create a flexible workforce.

In Chapter 6, the occupation of process engineer[34] is shown as the historical and organizational outcome of Taylor's strategy to separate planning from execution and place the planning activities in the hands of 'scientific experts'. Their responsibility is to determine the best and most effective method of part production and assembly. This separation is most clearly evident in NC programming which requires knowledge of a highly technical programming language and which is contractually methods' work. Many of the operations associated with APT (the machine tool programming language) are formalized with CAD/CAM through the use of software packages.

The separation of planning and execution is challenged with the introduction of advanced CNC equipment. The elimination of the need for technical programming skills and the 'user friendly' programming format or MDI (manual data input) merges these processes. Management at UFC, aware of the significance of this capability for the organization of work in the shop and the division of labour between machinists and methods, redefined this state-of-the-art technology as a 'conventional' machine tool. Local 8's arbitration case did not uphold methods' claim to the work, except in the case of stored tapes.

This manoeuvre had two primary consequences: (1) it eliminated methods from access to programming the machine tool; and (2) while it gave control back to the machinists, foremen could then use lower paid, less senior machinists to run it, removing the shop steward's discretion to negotiate over the assignment of workers.

Chapters 7 and 8 look at the impact of automation on the design process – drafters and design drafters and the impact of CAD on skills. Drafters, hired at UFC with little or no experience, learn necessary board skills on the job. Because of this reliance on company training, drafters' work – and the skills they need to acquire – are threatened by CAD. Moreover, drafters gain valuable experience doing basic design work on the board which prepares them for promotion into design drafting. As an information system, CAD provides detailed images along with design information, jeopardizing both the work drafters have traditionally been responsible for and their chances to gain valuable design experience. This increases the possibility that the remaining drafting work can be fragmented and/or routinized and performed by drafting clerks. However, the inevitability of the de-skilling process is mitigated by two factors: (1) the parts produced at Pine Hill are quite complex and cannot easily be reduced to routine procedures; and (2) the union is monitoring any shifts of work outside of the bargaining unit.

On the other hand, designers exercise considerable control over the use of their skills on CAD. Because they already have extensive design skills and have been working on CAD since its experimental stages, designers transformed the occupation from a manual to a computerized occupation over which they exert considerable control. Management has attempted to systematize CAD procedures, though with little success and no cooperation from designers. Nevertheless, the automation and simplification of some labour-intensive work, along with the availability of design information and the detailed images it creates, pose new questions about the meaning of 'creative' and 'non-creative' tasks. Such distinctions threaten the control designers have over the work and the demarcations between design and engineering. However, these changes have had little impact on the considerable skill design drafters are required to use in creating the design itself.

Part II of the book, entitled, 'The New Industrial Relations and QWL', examines organizational changes underway which are designed to increase flexibility in the production process and gain workers' cooperation in that process. Chapter 9 is an historical overview of management's efforts in this regard and a review of the relevant literature on QCs as part of this process. Many of the justifications for the implementation of programmes like QC have been based on arguments for industrial democracy. Proponents claim that such efforts will expand workers' participation in decision-making and lead to greater job satisfaction. Opponents argue that these changes have little to do with democratic participation and, instead, reflect an attempt to induce workers to cooperate with managerial goals.

QCs are part of the bureaucratic structure to introduce ideas for increasing productivity and efficiency and gain workers' cooperation in this process. The emphasis on the workers themselves generating these ideas (or more often, convincing them to adopt management's productivity measures) supports the organization's efforts toward increased flexibility.

Chapter 10 examines how management attempted to carry this out and workers' reactions to the programme. The workers who participated did so believing management's propaganda about improving the quality of working life. Resistance by managers to instituting workers' suggestions, their attempts to introduce topics more closely associated with their own goals and to control information from the meetings contributed to workers' declining interest and the programme's ultimate failure.

Chapter 11 presents the study's conclusions as concerning effects of automation and new industrial relations policies on the labour process. Computerization does eliminate manual work and the direct application of manual skills that have defined such occupations as machining and drafting. However, the knowledge and experience associated with practicing these skills are transformed into abstract processes. Computerized equipment does not, in most cases, replace traditional knowledge. Rather, it is an additional tool for use in problem-solving. It does, however, affect the work available as tasks are automated. Its effects on skills are complex and multi-dimensional and do not simply de-skill or re-skill occupations. These findings suggest that the analysis of the labour process needs to be reconstituted to include the different tendencies which emerge as both skills *and* tasks are transformed by technical change and management's implementation of that technology.

Quality Circles decentralize those aspects of decision-making which affect production and centralize long-range planning. Rather than being democratic processes, QCs are designed to address local management's concerns and create an environment for introducing change at the point of production. Whether these particular bureaucratic forms succeed or fail, they are part of the current ongoing process of corporate restructuring. The global economic conditions which contribute to these changes also lay the groundwork for new strategies for labour.

2 Site Description and Methodology

UFC's Pine Hill Plant was built at the beginning of World War II to produce airplane parts and accessories. At that time, the machine and production shops were very large with over 1000 machinists alone working at the plant. The United Electrical Workers (UE) organized the workers at the plant and represented them until a split within the union occurred and the International Union of Machine, Radio and Technical Workers (IUE) was voted as the sole representative of the shop workers.

During the 1950s, the shift from mechanical to electronic parts and equipment reduced the size of the shop dramatically. This was also the time when NC equipment was being introduced in the shop, reducing, to some degree, the numbers of machinists needed to maintain output levels. Currently, the Hurco has been added to the machine tools used at UFC. At present, there are approximately 100 machinists on site.

The shop is located in one section of the facility and includes the machine and sheet metal and assembly areas. In adjacent sections are the CAD room and the design and drafting areas. Engineers work in several sections, one of which is a large area sectioned off into individual work spaces with computer terminals at each desk. At the opposite end of the plant are areas which include test laboratories and the advanced computer facilities.

The four occupations chosen for the study – machinists, process engineers, drafters and design drafters – represent the production and design phases of work at the Pine Hill plant. Each occupation is responsible for working on a different aspect of the same part. These four occupations were studied to assess the impact of computerization on skills and, as processes become automated, the challenge to traditional boundaries between occupations.

The workers range in age from early 20s through late 50s. Some older workers have been at UFC for thirty years since graduating from high school or technical school and plan to retire from the firm. Others with less seniority view the job as a place to gain valuable skills and experience; they hope to move on to other firms. Yet other less senior workers who have family responsibilities would like to remain working at the plant.

METHODOLOGICAL ISSUES

This is an exploratory work on both the effects of automation on skills in related, yet very different occupations, and the nature and levels of labour-management conflict and cooperation within QCs. Therefore, a case study approach using in-depth, open-ended interviews was chosen to identify the range of processes which have an impact on the changes in the organization and control of work. Relevant union documents related to these issues – contracts, legal briefs and general records, as well as company newspapers and memos are secondary data sources which serve to verify statements made in interviews and provide additional information.

Until now, the impact of automation has been discussed in terms of either de-skilling or re-skilling. Little has been done in order to further understanding of the specific ways in which occupations have been affected. Moreover, analyses of QCs have tended to either support or negate its role in labour relations with few examinations of the actual content of such programmes or of the processes which define participation and outcome.

This research is based on a Marxian framework that focuses on the importance of a number of factors: (1) the social relations of production, including wider social and economic conditions along with particular organizational strategies, and taking into account divisions of labour, hierarchical forms of decision-making and specific bureaucratic rules and procedures; (2) the inevitability of class struggle and conflict as part of ongoing changes in social relations; and (3) an historical perspective on the changes taking place within the organization.[1] Heydebrand identifies these activities and outcomes as historically mediated. He states:

> this means that activity and outcome are *not* synchronic or simultaneous events, *nor* that they are immediately or mechanically related, *nor* that they are simple dualistic or logical opposites. Rather, outcomes may be seen as more or less incomplete, more or less imperfect historical objectifications of conscious, practical activity.[2]

Conducting the Research

Since I was unable to gain access to the plant either to observe and compare the use of conventional and automated equipment or to attend QC meetings to assess the dynamics of the process, a questionnaire had to be constructed to probe the issues of skill usage, changes and shifts in the lines of demarcation between occupations, and the processes of decision-making and labour-management relations within QCs. Open-ended questions were designed to elicit a broad spectrum of experiences which were subject to probes at any point in the interview process (see Appendix 1).

Focused interviews were conducted within an unstructured format. Respondents were asked to discuss specific experiences about working with conventional and automated equipment and participation in quality circles. Within this framework, the process itself was fairly non-directive. Respondents approached the questions from their own points of reference – what *they* considered to be the most critical features of automated work and participation in QC.

Site Selection

The choice of an appropriate site for this case study had several pre-conditions. First, the plant had to have recently installed computerized automation, giving workers time to become proficient with the equipment, yet able to compare this work with conventional tools and techniques and skill usage. Workers had to be able to discuss the differences in their approaches to a job on these technologies and accurately construct descriptions of the skills they used.

Second, the corporation needed to have an established QC program in operation for a minimum of one year. This would ensure that an accurate assessment of its 'life cycle' could be carried out and both its successes and failures analyzed. Since one of the aims of this work was to study the level of conflict and/or cooperation within QCs, a newly implemented programme would not provide the range of experiences of a more 'mature' programme and, possibly, reflect a 'Hawthorne effect' often associated with such organizational changes.

I decided to approach a labour union for assistance in locating such a corporation and gaining access to and the confidence of workers. I contacted the IUE for such assistance for an extremely practical reason: I am the daughter of a retired tool and die maker who is a member of the IUE. With no other, more formal, contacts, this introduction proved successful.

When the IUE agreed to participate, the choice of a site was the next step. The possible locations were extremely limited. Many plants had been closed or were in the process of closing – either relocating to areas with cheaper, non-union labour or simply shutting down as a result of mergers or corporate restructuring. UFC's plant at Pine Hill emerged as the only facility which both met the criteria for the project and was not threatened by closure, which would surely have an impact on workers' attitudes about participation and cooperation.

Four different IUE locals represent all of the shop, technical and engineering workers at Pine Hill. Only the clerical workers are unorganized. My original and primary contact with all the locals was through the union's International Representative. His introduction to the leadership of each of the locals and intervention at crucial points in the data collection process were critical to the completion of this project. He acted as liaison throughout the year and a half of negotiations with the locals and interviews with its members. At union meetings and functions, he expanded my numbers of contacts inside the union by introducing me to many of the stewards and members who he knew had an interest in the issues I was studying. In addition, his experiences as part of the original union organizing drive at UFC provided both an historical perspective on labour relations and changes in the production process at the plant since the period after World War II. Through him, I received the initial cooperation of the leadership of the locals.

Sampling Procedures

Each of the locals worked with me to identify appropriate respondents. The original sampling technique specified stratified sampling with age, years of experience and seniority at UFC as independent variables. Because the nature of this project includes understanding experience with automation and its impact of skills, it was considered important that there be a representative sample of older, more experienced workers, as well as those with less seniority. Experience working with computerized technology and perceptions of changes in skill may vary with age, experience and seniority. This sampling procedure had to be modified, however, after receiving a profile of the job qualifications for different occupations. Machinists and process engineers were only hired if they had experience in the field. In fact, a *B* machinist (lower classification) at UFC is equivalent to an *A* machinist (a higher classification) in other firms. Design drafters are promoted from within the company. Drafters are the only workers hired with little or no prior experience in drafting. Thus, while the sampling procedure continued to use age, experience and seniority as variables, it was also sensitive to the specific conditions surrounding each occupation.

Forty-nine interviews were conducted, including information interviews with five union officials. The breakdown of respondents is as follows: fifteen machinists; five process engineers; six drafters; eight design drafters;

and three sheet metal workers.[3] Six additional interviews were conducted with other shop workers who participated in the QC programme expanding the sample in this case in order to broaden the range of workers' experiences with QCs.

Among the respondents, the most senior workers were design drafters, ranging from eight to thirty-two years' seniority at UFC, with an average of twenty-eight years. The promotional ladder into the occupation from drafting accounts for such high tenure rates. Although machinists in the sample have an average of eight years' seniority at UFC, many have as much as twenty years' experience in the trade prior to working there. Process engineers average six years (although the population as a whole is more senior) and drafters average approximately four years. Since they are usually hired out of school and often seek promotions as design drafters, seniority levels tend to be low.

QC participants (sixteen) include current and former participants and represent almost half of all the respondents interviewed, excluding the machinists whose department was not chosen to participate in the QC programme. Three machinists claimed that they might participate if given the opportunity, but the rest were solidly opposed to the idea of labour participating in these programmes.

Of the remaining respondents, seven indicated receptivity, but were either on the night shift or, in the case of drafters, in work areas that wanted only design drafter participation. Six were completely opposed to the idea of any form of labour-management cooperation program.

Gaining Access

Each local provided computer printouts of their membership according to seniority and occupation. After reviewing the lists for possible respondents and before beginning the selection process, I held interviews first with shop stewards. Besides being contacts to workers, their experience as union representatives, knowledge of the union contract, and as workers in their respective occupations provided a valuable source of information on shopfloor practices and union-management conflicts. Their information added another dimension for understanding the problems facing both the union leadership and the rank and file around automation and QC and the issues of control, cooperation and resistance. The stewards were also instrumental in explaining the project to the membership and soliciting the cooperation of those who were chosen in the sample.

Early in the project, it became clear that gaining access to workers at UFC would be problematic. Management had flatly refused to provide me with access to any part of the plant and would not grant interviews. This denied me both the visibility necessary to gain workers' confidence as well as the opportunity to observe and question them about some of the more technical aspects of operating the equipment. Although the manager responsible for coordinating the QC program (and ensuring its successful operation) expressed an interest in the study and was willing to grant me an interview and give me access to QC meetings, because UFC had refused to participate, she felt she could not disobey company policy, not even to be interviewed after working hours.

I did manage to gain access to the plant on two occasions. The first was an 'Open House' for employees and their families. One of the shop stewards escorted me in as his 'spouse' and I had the opportunity to visit the machine shop, drafting and engineering areas in the plant and talk with workers who were demonstrating the automated equipment. The second opportunity came when the president of Local 3 brought me into the plant as his 'guest'. Before going in he said: 'This is really against the rules. Just don't cause a work stoppage. Then we'll all be in trouble.' This time I was able to speak with workers in the conventional NC and CNC areas and get a sense of the different ways skills are applied to the machining process.

Perhaps at this point the issues of gender and class should be raised. Overwhelmingly, the participants in this project are male and most are from working class backgrounds. There are very few women machinists, methods, sheet metal and other shop workers and, outside of the occupation, even fewer females who know much about the machine shop, particularly academics.

I had anticipated that my 'double' identity as female and an academic researcher could pose a barrier to participation by some who might feel awkward explaining exactly what it is they *do* on the job. I attempted to enlist the participation of some of the shop workers at a union meeting and found many to be extremely hesitant. When I asked their shop steward if there was a problem in my approach to them, he replied: 'You gotta understand. *No one* ever asks them about their work. They come to work, do their job and go home. Their supervisors aren't even interested in what they think about what they do. Along comes this person – a *woman* on top of that! – who says she's interested in what they do. They don't know what to make of it. They really don't know how to react.'

At that point, I decided that the best approach would be to emphasize our common class backgrounds and my father's trade as a tool and die maker. Growing up in a household where shopfloor culture was discussed, having family friends who worked alongside him in the shop, and memories of visits to the factory on 'family day', gave me an understanding of the way work is carried out as well as knowledge of workers' views of the job, of management, and their attempts to retain control on the floor and at the bargaining table.

This strategy proved to be successful. Some of the participants commented that knowing I am the daughter of a skilled tradesman made it easier to discuss issues of craft identity, work, skills and shopfloor conflicts. As one machinist said, 'Most people don't have a clue about what we do inside or what a machinist *is*, for that matter. I think it would be impossible for me to explain it to somebody who was completely from the outside.'

Since management denied me access to the plant premises, I met interviewees at the plant gate. My continued presence there provided additional opportunities for meeting workers. Men I had interviewed months earlier would stop by and ask about the study or joke about my 'unpaid career' as a graduate student. Sometimes, they would mention others who expressed an interest in being interviewed or they would introduce me to someone whom they thought might fit the profile of a potential participant. A few approached me to volunteer, explaining that, they had been they had been hesitant to get involved in the beginning. My visibility at the gate reassured them that I was committed to the work and demonstrated that their work mates were clearly willing to trust me.

In the following chapters, the voices of these workers provide an understanding of the ways in which decisions regarding the use of their skills and knowledge of the production process are organizationally mediated. At the same time, their experiences reveal the means by which these organizational structures are contested, giving rise to new strategies workers use to maintain control over their skills.

Part I
Automation and
the Labour Process

3 Review of the Literature

The purpose of this work is to analyze the technical and organizational impact of automation on job skills and the degree to which they are upgraded, downgraded and computerized, as well as the extent to which work is shifted between occupations. Automation, collective bargaining agreements and labour relations and control over the labour process are key issues which are explored in analyzing these changes. In order to assess specific changes in skill, data are presented separately for each occupation, describing, analyzing and comparing conventional skills and those used with automation in its various forms. Comparisons are also made between occupations to understand the differences in impact of automation on skills and those factors that affect control over these skills in production.

Recent research has generated a number of different positions of what changes in technology might yield. At the extremes, computerized automation is presented as either the source of freedom for human potential and creativity, or the ultimate source of domination over an increasingly monitored workforce. A key issue is the debate over the de-skilling or upgrading of affected occupations.

Researchers have taken three basic approaches to the problem: (1) the emphasis on the role of managerial control in determining how technology is used in the labour process; (2) the importance of wider market and organizational forces in determining managerial strategies for restructuring the technical and social divisions of labour; and (3) technology as the primary factor influencing job skills.

MANAGERIAL CONTROL AND THE LABOUR PROCESS

The debate over new technology and its de-skilling/re-skilling effects has focused primarily on Braverman, who argues that long-term tendencies in the labour process are toward de-skilling and the degradation of work and emphasizes the central role of managerial control in the reorganization of the labour process. He notes:

> It thus becomes essential for the capitalist that control over the labor process pass from the hands of the workers into his own. This transition

27

presents itself in history as the *progressive alienation of the process of production* from the worker; to the capitalist, it presents itself as the problem of *management*. [1]

According to Braverman's thesis, Taylorism, or scientific management, has been the key feature in the devaluation and dequalification of work. For years Taylor studied machine work, identifying twelve independent variables that had to be considered in planning and executing the machining process. [2] He proposed the application of time study to these variables defined by two broad categories: analytical and constructive work. The first represents the identification, recording and indexing of all elementary movements, the study of working time for each movement and the selection of the best way to perform them. The second refers to the reorganization of the work process in accordance with the above findings, including the standardization of machines, tools and work conditions. Although Taylor was not successful in devising a plan for the separation of these variables into discrete tasks, he did manage to discover the use of high-speed steel for running machines faster. His recommendations for the standardization of tool shapes and speed and feed calculations for running these tools were adopted industry-wide. Those elements of planning were removed from the shopfloor and the control of the machinist and became a function of the planning department.

The fragmentation of jobs, the increasing homogenization of skills, the proliferation of narrow and specialized skills, repetitive work and wage payment systems are all the result of Taylor's recommendations for scientific management.

Earlier works in sociology on the labour process focused on the 'plant', [3] with in-depth analyses of subjective and objective dimensions of 'life on the line' in specific industries. Later work [4] compared the effects of different types of automation on the organization of labour and the skill and autonomy of the workforce. Braverman refocused this analysis within a much broader framework of the transformation of work under advanced capitalism and managerial efforts to organize and control the labour process.

One of the weaknesses of this thesis, however, is the over-emphasis on the de-skilling process. Braverman presents Taylorism as ultimately successful in breaking down the resistance to the reorganization of work by technical and/or bureaucratic means. However, as Taylor himself discovered, scientific management was limited in the effects it could have on some complex skills.

Braverman's focus on the objective conditions of the capitalist labour process has been challenged by a number of other researchers. Burawoy[5] states that the subjective elements of work must be incorporated into an analysis of control which Braverman identifies as the distinguishing feature of capitalist social relations. He asserts that there is a constant tension around the separation of conception from execution – 'too little separation threatens to make surplus transparent while too much separation threatens the securing of surplus.'[6] Manwaring and Wood focus on the working knowledge applied in the labour process which gives workers a measure of control over day-to-day activities, even in jobs which are identified as 'unskilled'. They state:

> Working knowledge does not in and of itself refute the de-skilling thesis, but it does provide a different vantage point, one in which the central notion is that work is both degrading and constructive, both crippling and enriching.[7]

Workers' discretion continues to be a critical element, one through which they maintain an active role in the labour process.

MARX AND THEORIES OF THE LABOUR PROCESS

Using a Marxian analysis, current theories on the labour process integrate technology and managerial control into a broader framework of existing social relations. Marx[8] discusses the relationship between the forces of production (technology and materials) and the relations of production (the organization of work) as mutually conditioning processes. Technology affects and embodies social relations as well as being affected by them.

Thompson focuses on the importance of existing social relations and the forces that affect these processes and which are, in turn, affected by them. He develops this idea in his analysis of the application of technology to the labour process.

> In contrast to the conventional vision of a neutral technology determining the nature of production, its social construction is located inside class relations and their antagonism. The capitalist labour process is therefore subject to a number of identifiable tendencies, whose central features are de-skilling, fragmentation of tasks, hierarchical organization, the divisions between manual and mental work, and the struggle to establish the most effective means to control labour. It must be stressed, however, that they are trends and not finished

processes. Each aspect can take a variety of historically relative forms.[9]

As trends, they embody contradictory tendencies. The nature of de-skilling, the division of labour and the organization of work are all subject to the specific demands of production as well as to the struggle between labour and capital to exert control over this process. According to Thompson's analysis, the impact of technology on skills is neither predetermined nor uniform. Child[10] builds on this framework by placing these processes within the context of managerial strategies. His criticism of Braverman's position is that controlling the labour process is only one strategy. His representation begins (as Braverman does) with the fundamental objectives of capitalism (what he terms 'basic strategic motives') leading to such traditional managerial objectives 'corporate steering devices' as cost reduction, increased flexibility and enhancement of control.

However, Child emphasizes the critical importance of other intervening variables which give particular formation to the use of skills and the adoption and implementation of technology: product and labour markets, availability of technology and workers' and unions' responses. According to Child, each one has implications for the use of technology in the labour process and its effects on skills.

THE TECHNOLOGY DEBATE

There are two opposing positions on the effects of technology on skills. The first, presented in the works of Noble[11] and Shaiken,[12] views technology as a real threat to traditional craft work and the cause of a drastic reduction in workers' control over the labour process. The second perspective, as developed by Adler[13] and Hirschhorn,[14] sees the re-skilling potential and the creation of technically oriented and upgraded jobs as a result of automation.

Both Noble and Shaiken focus on the decimation of manual machinists' skills, which are the historical trademark of the craft, and the removal of much of the decision-making and planning and the exercise of skill from the workers. Noble concentrates on the historical role of management and the government in influencing the design of machine tools which would marginalize the control machinists have over production and minimize the need for highly skilled labour.[15]

In the 1950s, NC was touted as the step toward the automated factory by trade and business publications.[16] According to their predictions, the tireless machine – driven not by recalcitrant machinists, but by automatic instructions – would solve problems of productivity and labour relations. Ultimate control would finally rest in the hands of management. *Business Week*'s cover[17] represents the shop manager's dream with its caption, 'Machine Tools that read blueprints'. But does it, in fact, accurately reflect conditions on the shopfloor? According to Noble, even as trade journals lauded the introduction of NC, there were others in the industry (mainly machine tool manufacturers) who warned of the false hope of doing away with skilled labour.[18]

Noble's work does, in fact, confirm the non-technical, social decisions surrounding the split between planning and execution. In addition, he reveals the contradiction between the propaganda that industry analysts promote on the effectiveness of automation and the shopfloor experiences of users of these systems.

Noble also counters his own argument that the complete destruction of skills is built into NC. He cites a study conducted by the US Machine Tool Task Force indicating that the formal requirements of NC work run counter to the variability which exists throughout the machining process.[19] Essentially, this is the same problem F. W. Taylor had when he attempted to systematically break down the machining process into simplified tasks. The process consists of so many discretionary tasks – both in planning and execution – that Taylor was never able to complete the project.[20] The development of NC might have been an attempt to re-introduce Taylorist strategies through the automation of work processes and a division of labour based on technical programming skills and limited machining practices. Nevertheless, according to the data Noble presents, production needs also played a significant role in shaping the technology.

Through interviews with machinists and management, Shaiken documents specific de-skilling processes and the increased marginalization of machinists who work with automated machine tools. His first-hand knowledge of machining gives this work a more detailed view of the subjective experiences of the machinists whose jobs have been automated in addition to the different levels of skills that have been altered.

Shaiken concedes that automation cannot remove complete control from the shopfloor. Machinists are still needed to intervene when the technology is unable to accurately or efficiently perform certain operations. Moreover, he acknowledges that there are a range of managerial options available in applying automation to the labour process that does not necessarily mean de-skilling. According to Shaiken, it is the flexible nature of the technology that offers a promise of job enrichment or the threat of intensive de-skilling.

The other side of the technology debate focuses on the effects of automation on general categories of labour rather than on specific labour processes. Hirschhorn and Adler stress the potential for creating new categories of technically skilled jobs and re-skilling many of the occupations affected by automation.

Hirschhorn's work is based in socio-technical systems theory, which defines technology as the central feature in determining new forms of work design. Although he refers to 'social and political interests'[21] that shape the organizational design of work, Hirschhorn presents a basically technological determinist position. While he does refer to the processes of integration as tendencies, fragmentation, intensification and routinization of tasks are never addressed as possible outcomes of the same technology. The integration of work, flexible use of skills, and the emergence of work groups are, in his view, indicators of an organizational form determined primarily by the imperatives of the technology. Social relations become, instead, 'behaviours' that are distinctive to the technology. The definition and exercise of authority are described as part of the design of the system which fosters a more 'cooperative arrangement' between and among workers and management. His work lacks historical explanation for these relationships other than to say that they are a function of technical design. Wage payment systems were created, according to Hirschhorn, to coincide with the prevailing stage of mechanization and corresponding job structures. They are, in his view, part of the 'work culture', rather than a part of the systematization of managerial practices.

Adler claims that automation has, with 'the overall shift in the structure of the economy, . . . favored higher-skilled jobs'.[22] He questions the arguments on de-skilling, suggesting instead that new skills replace those removed by automation. A case in point is the machinists who may be re-trained as programmers. Although he claims this is a rather routine managerial practice, there are no data provided to indicate the extent to which employers have availed themselves of this resource within their firms. In Adler's view, this would be a 'smart' move on the part of management. 'Smart', however, is not always what determines the use of skills in production.

Acknowledging that not all machinists experience upgrading, Adler focuses on the importance that responsibility assumes in working with automated processes. Like Blauner,[23] he equates the practice of skill with acting as a 'responsible employee'. Skills that are no longer used because they have either been automated or fragmented are still necessary in order to monitor the system. Adler seems to equate the notion of acting responsibly and with a degree of autonomy with the direct exercise of skill.

While Adler is correct in stating that massive de-skilling is exaggerated, his treatment of skill as a uni-dimensional category distorts the role of 'responsibility' as a replacement of skills lost through automation and overstates the acquisition of technical knowledge as an indicator of re-skilling.

DEFINITIONS OF SKILL

Cockburn's[24] definition of skill provides a framework for identifying and analyzing the different components of skill that are affected by technological and organizational changes. As Cockburn states, there are several meanings of the word skill which need to be unraveled and distinguished from one another. According to Cockburn, skill consists of three dimensions:

(1) the skill that resides in the man himself, accumulated over time . . . ;
(2) . . . the skill demanded by the job – which may or may not match the skill in the worker . . . ; and (3) . . . the political definition of skill: that which a group of workers or a trade union can successfully defend against the challenge of employers and of other groups of workers.[25]

Each dimension expresses a different aspect of the organization and control over skills and the ways in which labour and management struggle over its content and use. Not only does this breakdown allow us to address the specific labour processes that are affected and altered, it also enables us to see how they are defined within the context of capitalist social relations.

In addition to distinguishing between different meanings of skill, Cockburn warns against over-simplifying the complex effects of changes on skill. She cautions 'about supposing that de-skilling for one group implies an overall de-skilling in the enterprise'.[26] In addition, Cockburn states that distinctions need to be made between 'loss of skill from loss of control'.[27]

The breakdown of skills into these elements provides useful categories for analyzing the trends Thompson outlines and the effects of techno-logical and organizational changes on these different components of skill. Together, they provide a comprehensive strategy for examining the precise forms these changes assume within the context of exist-ing social relations. This analysis is, simultaneously, an examination of changes in the organization of production and categories of work, as well as of the elements which comprise specific skilled labour processes.

A study conducted by the Office of Technology Assessment[28] on the application of programmable automation in metalworking industries provides a framework for analyzing shifts in the properties of skills which Cockburn develops. The report examines specific processes affected by the introduction of computerized technology, including: (1) task displacement – those tasks previously performed by people which have been transferred to machines; (2) task creation – maintenance and support work associated with automated equipment, in other words, programming; and (3) alteration of the types of skills required to perform the necessary operations. According to the OTA report, skill composition can be measured in terms of its 'depth' and 'breadth'. 'Skill depth refers to the input needed to perform an individual task or group of interconnected tasks . . . (and) has two dimensions: time to proficiency and judgment.'[29] It includes education and training, and the application of specific techniques to perform a job, along with the ability to exercise judgment in the use of these skills. It is, in fact, a definition of traditional craft skills. Skill breadth 'refers to the input needed to perform a set of (non-similar) tasks . . . (and) applies more to jobs or occupations than to single tasks.'[30] It is primarily concerned with the occupational categories and hierarchies within work environments affected by technology and managerial decisions to reorganize the structure of jobs within the corporation.

These concepts – task displacement, task creation, skill depth and skill breadth – are extremely useful for assessing the changes in specific skill requirements, the division of labour, and movements toward job redesign. They provide the means to analyze both the objective skills and subjective decision-making processes used in production, and a basis to examine the shifts in the organization of work through the re-definition of job classifications and the re-allocation of tasks.

By combining these different levels of analysis, the following chapters explore such questions as: under what conditions are jobs de-skilled and which particular aspects of skill are affected? Conversely, what conditions can and do lead to re-skilling and which processes are upgraded and/or acquired to reflect this shift? By examining changes in the social and technical organizations of work and analyzing which skills are affected and the ways in which they are used in the production process, the following chapters offer insight into both these questions.

4 Automation on the Shopfloor: Machinists

> There's a school of thought that says that we will go the way of the blacksmith and the cooper. I've been hearing things like that since I was in high school.
>
> *Machinist at UFC with 30 years experience*

CONVENTIONAL MACHINING

Machining has been described as 'part science and part magic'.[1] It is a craft not easily understood by strangers to the occupation and not easily described by those who practice it. The machine shop at UFC's Pine Hill plant is a complete floor of conventional machines which, at first glance, look like huge abstract industrial art pieces with levers, spindles and handles extending from their frames. The larger, numerical control machines are in a room separate from the rest of the machine shop and resemble conveyors of different sized and shaped tools which, on command from a computerized program, spin around, choose the correct tool and carry out drilling and milling processes. There are a few smaller and much quieter machines – whose trade name is Hurco. They are the most advanced CNC equipment UFC has purchased. The men operating them appear to be somewhat more relaxed than those who work in the NC room – although just as alert.

Metal chips removed from machined parts cover the floor, workbench and often the workers, giving an appearance of shiny grey dust on their clothes. The noise level is constant with the sounds of banging and pounding and the whine of metal grinding metal. The smell of the shop is an unmistakable and unforgettable pungent odour of oils, grease, and coolants.[2] In addition to the physical and mechanical work they are engaged in, machinists are alert and attentive to every sound and movement around them. Awareness of changes in the sound of the machines or erratic movement of the tools can mean the difference between a safe job and the loss of a finger or hand, or other serious bodily injury to themselves and their co-workers. Because of the precision and danger of their work, machinists display a mixture of pride and bravado as they describe the work they do. Only skilled machinists are hired at UFC; there are no training or apprenticeship programmes other than an initial trial period.

Early innovations in machine design at the turn of the century saw the emergence of more specialized equipment with elaborate jigs and fixtures. Essentially, the process of control remained on the shopfloor. At the same time, however, the move toward specialization narrowed the range of required skills. The need for specialized tooling expanded the role of the tool and die maker in the shop at the expense of the all-around machinist. This, however, left management with another shortcoming – the inflexibility of their equipment. The development of numerical control seemed to answer the problems of expensive and inflexible machining, restricted output and machinists' control over planning. Conventional machinery has dials, levers and cranks that machinists turn and adjust as they guide cutters through pieces of metal to produce parts. They work from a blueprint drawn up in the drafting department and often, but not always, from an additional operation sheet (OS) prepared by methods. In many smaller job shops methods, as a separate occupation, does not exist. The machinist methodizes the job, working directly from the print or a set of engineering instructions. As one machinist explained his previous experience in a small shop:

> This place I worked at before I came to UFC – the engineers would come down with sketches and prints and say, 'Listen, this is what we want. Can we do it?' And you work together with the engineer on putting models together and tryin' this and tryin' that. *That* was great! You'd work together with them and work on ideas. Now *that's* the ideal situation. I even drew up my own blueprints.

At UFC, machinists will work without an OS if none is available, disregard it if they believe the instructions are incorrect to produce an accurate part or know a way to do the job more simply and/or more safely.

Fundamental machining skills include a knowledge of algebra, geometry and, to some extent, trigonometry; blueprint reading; an understanding of metallurgy; how to grind and sharpen tools; and knowing how to check finished parts for their accuracy. In addition, a machinist must understand the capabilities of each machine and the calculations of appropriate speed and feed rates for that machine and the particular job. Speed rates refer to the speed of the cutter as it machines a part. This rate is based on the composition of the cutter (steel, carbide tungsten, etc.), as well as the type of metal being cut (aluminum, steel or alloy metals).

Feed rates determine the type of cut the machinist needs to make – the kind of chips to be taken off the metal. Small, delicate cuts produce a different chip in shape and thickness than do larger, deeper cuts. Both speed and feed rates are dependent on one another and must be taken into account when determining a job. However, these skills also involve the use of judgment and experience. It is a job which is not divided into discrete parts and requires careful planning. As machinists told me several times:

> You have to think of the operations that are gonna come ahead. You always have to think ahead.

> I ask myself, 'What is the closest dimension I'm gonna have here? How's that gonna affect the rest of the job?

> You have to be able to see the part before you plan your job.

The development of keen sensory perception is critical in this trade. The sound of the cutter, the 'feel' of the cutter which is altered using dials and manually adjusting the speed and feed as it goes through the metal, and the attention to the size and curl of the chips coming off of a piece are the types of sensory skills used in conjunction with the more 'objective' machining processes.

In their descriptions of conventional machining, workers discussed the critical role of sensory skills in the production process:

> A lot of it we do by eye and ear – [to determine] the sharpness of the cutter. By using them, we adjust the feeds and speeds to get the proper finish.

> Vibrations are always a problem that something's gone wrong. You hear it and you know right away. You can sense a dull tool from hearing. It's not always that obvious, but you can pick it up just by listening to the sound the cutter makes. I use my ears a lot. For threads [close tolerance work], I use my sight. You have to release and pull back exactly. You have to be quick.

> I think to be a machinist, you have to start off on the old ones [conventional machines] – if you're gonna be a *good machinist* . . . When you do it by hand, you start getting a feel for feeds and speeds to get your best finishes and you do that by turning a dial by hand . . . You learn a feel for coming up – tickling up to a part. It's something that you just learn.

You learn (by using your senses) whether maybe it's going a little too fast. The tool doesn't sound right, or the chip isn't coming out right. You can tell a lot by the way a chip comes off a piece of metal. How it curls. There's a certain speed and feed – you get the right combination and you're gonna get a nice, long curly chip. If the tool gets dull, the chip's gonna start breakin' up. The tool's gonna start wearin', it's gonna start smokin' and you're gonna get a lousy finish. You're goin' a little too fast, you're gonna burn the tool. If you're takin' too much metal in one cut – *ya know* – these are the things you just *know*. This is what comes together over the years.

The importance of judgment, discretion and experience in conventional machine work is apparent in these descriptions of sensory skills. Conventional machining requires that the machinist coordinate complex physical and mental processes and apply them in whatever appropriate form he believes will produce an accurate part. Sensory skills contribute to both the artistry of the work and basic skills of the trade. It was often difficult for the machinists to express, in concrete terms, exactly how these skills are applied. Individual sensory development merges with technical knowledge and expertise.

What underlies the use of both types of skill is the ability to plan and execute the work process. These machinists are quite aware of the control they exert over their work. The autonomy to plan the setup of a job and actively intervene in the production process to get a more perfect cut or finish on a part is, according to them, the definition of what a skilled machinist *is* and *does*. They measure their own skills and capabilities and those of other machinists against these criteria. Moreover, responsibility for the quality of the parts produced ultimately rests with them. Scrap pieces and broken tools (including those made while following operation sheet–OS–instructions) are often treated as machinists' failures, indicating the degree to which they are expected to exercise independent judgment and skill. The hallmark of a skilled machinist is the extent to which he is responsible for planning work and exercising control over the use of skills. This is evident in the following comments:

I love looking at the blueprints and saying, 'OK, now how're you gonna do this? I've done it this way before, now I'm gonna try something else.

They gave me an aluminum job to face on a milling machine and then go to the lathe and turn it over (OS instruction sheet). I said, 'Why

should I bother milling? Let me just try the whole thing on the lathe.'
It worked out excellent. I had some initial problems with vibrations, but
I worked it out.

For every guy in the shop – everybody does it a little differently. That
little thing gives you the creative impulse . . . Everything that you've
been able to learn comes together: which tool's gonna work better, using
different oils and different feeds and speeds, using different angles on
your tool bits for different metals. These are the things you learn over
the years. You're involved in the whole process – right down to the
littlest thing.

Unlike mass production facilities where the emphasis lies on repetition
and maximum output, small batch facilities, like UFC's Pine Hill plant,
depend upon skilled machinists' labour for high quality, precision parts.
These machinists are well aware of their capacity to exercise independent
judgment in planning the work process.

Their ability to control the planning and organization of conventional
work is a source of pride for these machinists which was evident throughout the interviews. They all point to the critical importance of their
discretionary use of skills in production and the freedom it provides to
refine their skills.

All of the machinists interviewed structured the descriptions of their
craft in terms of emotion or subjectivity – self-satisfaction, artistry, pride
or what can be defined as a craftworker's sense of identity. The following
comments are their own introductions to the issue of skill:

Machining is really a lot of intuitive knowledge. I guess you can say
it's a sixth sense.

To me, being a machinist is like being an artist. Like an artist who
takes a piece of wood and keeps cutting it until you see the face
coming, the hand, the hair and everything. Then you feel proud of
your work . . . You get so anxious to see the finished product. I even
went home one time and kept thinking about what I was going to do
the following day. That's what hooked me on being a machinist.

Conventional machining is 90% hard skills and the other 10% is pure
art form.

I enjoy a tough job – a real challenge. And then to finally see the
finished product . . . That's a real nice feeling . . . I like the lathe work
best – you get such nice finishes.

Conventional machine work incorporates all three dimensions of Cockburn's analysis of skill. Sensory skills, only gained by experience and developed through years of practice, are essential to the work process and are, by tradition and union agreement, part of the definition of the trade.

NUMERICAL CONTROL

Automated machine tools are based on programmed instructions that guide the machine through its cutting motion without the manual intervention of a machinist. Machining information is punched onto a tape using codes that represent the sequences to be performed. Each is a command which informs the machine to carry out particular metal cutting operations. However, as Shaiken points out, 'the machining process itself does not change', rather, NC is a 'means of information processing and machine control'.[3]

A central feature in the development of NC was the role of the programmer in the machining process. Planning the process, rather than the performance of skilled operations, became the critical element in NC work. The discretionary use of skills was contained and limited through the application of computerized instructions. It would appear, on the surface at least, that Frederick W. Taylor's dream of separating planning from performance had been realized.

The creation of a separate department to assume responsibility for technical planning marked the beginning of management's intrusion into the machine shop. Although most of the decision-making remained on the floor, aspects of planning were removed to a distinct group – the process engineers. Which tasks would remain on the floor became a source of tension and conflict as technological change was introduced and management reorganized the production process. Decisions to remove parts of the planning or to assign new technical planning methods away from the floor are neither uniform nor inevitable – they are both a conscious and arbitrary process on management's part. However, management's success at removing decision-making is contingent upon both the nature of production needs and rank and file resistance.

The removal of significant aspects of machining is a source of anger and resentment over what workers perceive as an infringement upon the most basic elements of their craft: their ability to use their skills and judgment. Moreover, it has also contributed to a decline in their sense of power, authority and control over the labour process. It is acknowledged both in the literature and among the machinists themselves that the degree of experience and expertise needed to operate NC equipment is much below that of an experienced machinist.

Since NC eliminates virtually all of the physical operations – and the level of control that is exerted over these processes – the tendency of researchers is to claim that *all* skills are eliminated. Conventional work depends on the active intervention of the machinist at various stages of the production process. The mechanization of much of the physical work and the computerization of the sequences of operations generated by tape-controlled methods eliminates these traditional forms of planning and control. Direct access to both the planning and production process is mediated by the technology and creates a more rigid division of labour between machinists and methods. Working on conventional equipment, machinists can produce parts without an OS or revise the instructions. NC machines operate from a specific set of programmed instructions which are not as easily altered.

Even as the machinists are extremely critical of the encroachment of NC on conventional skills, they acknowledge that there is an increased emphasis on conceptual skills and problem-solving. Specifically, they talk about the ways they apply their skills to NC setup work.

> You can be less of a machinist on automatic equipment to make it run. To set it up – no. You have to be more of a machinist. You've gotta know if a goddamn tool is gonna cut. You've got to program the feed an speed. You gotta make damn sure that your tools are right – they're in the right position, they're where they're supposed to be at the particular time. And you have to know when a tool is not gonna work because the automatic machine has no conscience.

> The setup on an NC can be more sophisticated. So, I guess some skill levels rise – to a degree. You have to do a number of operations all at once – you have to see the finished product. Manually, you do it one at a time. If I'm setting up a job on NC equipment, I might do a diameter or perimeter, and drill some holes. On manual equipment, I'd have to do the perimeter first, then do the drilling. Now, on NC, I set up a number of tools at the same time. So that thought process is a bit more sophisticated [on setups].

> You can be creative on NC with the setup. You know any job can be done a thousand different ways. So, you might work together [with the lead man] to come up with something that's totally different. You still have to go through the setting up process. The only thing that's on the tape is that the machine has to start from position X. You gotta

make sure you're in the exact spot where it tells you to be because
the machine has no idea of how to get there. If there's bends in it, the
machine don't know that. So everything is not gonna be to tolerance.
You gotta know how to clamp it down. You get rid of the bends. If you
don't know that, well . . . On NC you just have to figure out how to get
it to this point, then it's the machine's responsibility. If you don't know
how to do that, than you're lost and the machine's lost.

These experiences contradict the positions that NC de-skills machinists'
work. Critical skills remain important and machinists are expected to
adapt them to NC work. The automation of machining processes requires
a greater effort in coordinating the different conceptual steps to facilitate
the machine's programmed movements. Altlhough NC has fragmented the
labour process by taking programming off the floor, the remaining conceptual work requires skilled machinists to plan and set up the process.

Machinists continue to exert some control over automated production
work. Planning a setup includes assessing the impact of each step on the
procedure following and on the entire job without the input of sensory
skills. It is, in fact, a different level of conceptualization – an increasingly
abstract process which requires the application of skills and experience to
a more limited, yet critical, aspect of machining. All of the variables that
can affect the process must be accounted for in planning the setup.

In addition to the increased importance of conceptual skills, machinists
cite the increased flexibility of NC equipment which enables them to work
on setups for jobs that are impossible to do on conventional equipment. As
several commented:

> On NC – the type of parts they're capable of making – the conven-
> tional machinist would never have the opportunity to even see these
> parts being made, never mind set up a machine that's making them.
> They just couldn't do these things on conventional machines.

> On NC, you can make and design parts and put together units that was
> impossible forty or fifty years ago. You can generate some radiuses on
> conventional equipment, but it's so time consuming. We have whole
> new parameters open. On certain jobs, where you might have a hundred
> to a hundred fifty check points – on what they call inkadoos that they
> run on the K&T machines. They might have as many as two hundred,
> three hundred holes. They're thin-walled housings. You can't do it on
> conventional equipment. It actually opened up new parameters as far as
> design and utilization of tools.

The increased capabilities of the equipment and the sophistication of the NC setups require a machinist to translate the technical and sensory knowledge used manually through a conceptual format.

Control over planning the operations becomes a contradictory process for machinists working on NC. On the one hand, their ability to apply a set of craft skills to a conceptual process is an indication of their indispensability in automated machining. On the other hand, the span of control they are able to exert is circumscribed by the tape-controlled planning of the methods department. Machinists remain central to the process, yet limited to specific segments of NC work.

However, control over the rest of the process is not entirely eliminated. Machinists continue to intervene in several ways: in the initial 'prove outs' of NC tapes, in the use of overrides, and in their ability to edit and program tapes. Once the machine has been set up, the machinist monitors and maintains the process as it has been programmed.

Prove-Outs

Before a job is allowed to run, the tape must be 'proved out' to remove any errors in the program. At UFC, this is the responsibility of senior (AA) machinists. 'Prove outs' are an interesting aspect of the job. They involve running through the program and assessing the accuracy of methods' work. No part is actually produced. The machine 'cuts air' as the machinist monitors the process and checks it for accuracy. As the machinists describe it:

> The first time you'll run it through very slow to see if this guy has done his job right. Most of the time he hasn't. You use the readout [while the machine is running]. It will tell you exactly where the machine is going to go. One of the biggest mistakes is that the math is wrong . . . The machinist sits there and looks at the program and knows that the machine can't do what's been programmed because it will bomb. So, then you call methods and they come down and change it.

In effect, the machinist *approves* methods' work. Their judgment about the accuracy of the procedures is the basis for 'prove outs' in the production process. These experienced machinists are expected to use their skills to evaluate a program designed off the shopfloor. It is an indication of the degree to which skilled machinists remain a vital part of automated work. As one machinist remarked, 'We can do a job without methods, but they can't do the job without us'. However, the work of planning a job is, under the union contract, no longer their province. However, as Chapter 5 will show, the next generation of computerized machine tools challenges the contractual separation of planning and execution.

Monitor/Maintenance

Once a job has been 'proved out', it can be run by any machinist. It is here that the decline in skill is identified and most often cited by critics of NC technology. The simplification of machine operations through automation and the division of labour between planning and execution gives management a range of options with which to organize the labour process. In many instances, NC provides the means to develop two classes of machine shop workers: the skilled machinist who, in some cases, programs and runs the equipment or sets up and 'proves out' the job and solves production problems; and the unskilled or semi-skilled operator who monitors the job as it is being machined and who simply starts and stops the process. Although a need remains for highly skilled machinists, shortages exist. The shortages are as much a response to the increased use of automation as they are an indication of management's resistance to training and hiring skilled machinists.[4] Moreover, increases in foreign competition and in the diversity of products and materials used in the production process also affect employment levels. Although NC machinery comprises only 4.7% of all machine tools in the US,[5] its effect on the industry is increasing. Productivity rates are higher in part because the machines' movements are pre-set and performed virtually without interruption. As more of the older conventional machines are replaced by NC and CNC, further increases in productivity can be expected.

For the machinist, once the setup has been completed, NC work involves primarily the monitoring and maintenance of the process as it has been programmed. The machinist receives a readout of all the operations to be performed and uses it to judge the accuracy of the machine's performance and to anticipate upcoming machining sequences. The readout is intended to compensate for the loss of some elements of control the machinist had in planning conventional work. The machinists comment on the ways in which they monitor the process:

> In order for a machinist to be a *good* machinist on NC he would have to know how to read and interpret the readout. He can't handle problems that come up without that tool. If that's not available to him, he's kinda workin' half blind. You have the CRT to watch as the part is being made and you also have the hard copy with everything on it. So, if you wanna go down 30 tools and pick up something, you can.

Keeping up with skills (on NC) depends, to an extent, on the machinist. You have to sit and read the book and learn what the machine does. They make a tape with all holes and they stick it in the machine and it gives you a big readout of each step. You gotta sit down and read that – what each tool is doing, where it's going, learn what all the codes are for. You're increasing your skills. Just by doing that, you'll learn how to put in a program yourself. If you learn how to read the readout, you could learn how to program on NC.

Readouts provide enough information to allow machinists to understand and intervene, when necessary, in that process despite the elimination of traditional methods of planning and control. Without such a tool, scrap and injury rates and machine damage would be excessive. For an experienced, skilled machinist readouts provide a mechanism for continued control over aspects of planning done both off of the shopfloor and within the process itself. However, the interpretation of this information does not constitute a new skill, nor does it replace conventional means of control and decision-making. Those processes have, to a great extent, been automated and translated into NC programming codes. Nevertheless, readouts act as a wedge for machinists to remain involved in the process, and as a means to gain access to NC programming in order to make minor adjustments in the program.

The job of monitoring and maintaining NC is not an entirely different set of experiences from conventional work. As Tulin notes:

Many of the ornery, persistent problems of metalworking are the same on NC as they always were before, in manual work. Drills run off course. End mills walk. Machines creep. Seemingly rigid metal castings become elastic when clamped to be cut, and spring back when released so that a flat cut becomes curved, and holes bored precisely on location move somewhere else after they're made.[6]

Overrides and Edits

The process by which alterations in the program are done to correct errors or machine failures is through the use of the override and edits. The override is a switch on the machine that allows the operator to adjust the feed and speed rates without having to re-program the tape off the floor. Machinists can use it to lower the rates if they believe it will damage the tool, prevent an accident, or just to stretch out a job.

Some of the machinists view the use of the override as a means to insure that key decisions will remain under their control. They see it as their 'insurance policy' against becoming simply operators of NC equipment. As a few commented:

> There's a chart with feeds and speeds that these tool companies put out. To me, the people who make those tools put out those charts at too rapid a pace because they're in the business of selling tools. We have a guy upstairs in methods who swears by these charts and he burns out tools and ruins parts and gets lousy finishes. Being a *machinist*, I just won't do that.

> On those NC machines in there, you inherit somebody else's program [from methods]. Now, the man might have been a machinist. Most of them up there work off of a book format – in other words, you take the same size cut, it don't make any difference how you're holding the part, or what the material is – whether the tools were made by a foreign concern or you're working with prime stuff. You're strictly at the mercy of somebody else. The override gives you some control over some of the more stupid mistakes that are made by methods.

> I never use an override. I use it sometimes to tune a tool and what feeds and speeds I have I put in the program. I have a guy comin' in the next morning and he sees that something's programmed a certain way, he's got to expect that to be in the machine. Sometimes you'd use an override and forget it and you leave it on one tool and the doggone thing will kick in the wrong setting and somebody might get hurt.

> When something is wrong, I just override the tape and try to figure it out for myself. I usually don't bother calling methods down to the floor.

The ability to use the override for 'prove outs' and adjustments indicates the extent to which machinists at UFC are able to influence the automated machining process. Overrides are expected to be used in 'prove outs' to work with methods in the creation of a more accurate tape. Yet, overrides are contractually off-limits to machinists. Once machining sequences are taped, the process is considered methods work. The formalization and computerization of planning redefines it as a technical process. Given the fact that, in this shop, the division of labour between planning and execution had already existed, programming the operations seemed a 'logical' extension of methodizing a job.

The override allows the machinist to deal with the uncertainty that arises in the course of production. Adjustments that have to be made because of a machine's idiosyncratic tape interpretations or simply to modify an operation to get a better finish are difficult to anticipate beforehand. Without this level of discretion, safety and precision work are impossible to maintain, scrap rates would rise and work schedules would not be met.

As the data indicate, NC has several major consequences for the work of skilled machinists. It affects: (1) the nature of the planning functions that remain on the floor and those that are removed; (2) the loss of control over all physical work related to traditional skills; (3) the increase of uncertainty due to the loss of control; and (4) methods of insuring or increasing control with the acquisition of new skills and the application of traditional skills in different forms.

Another process machinists use to exert control in NC work is tape edits. For an experienced, motivated machinist, there are several avenues open to learn this skill. At UFC, some of the men learn it through self-study by taking the manuals home and reading them. Others enroll in NC programming courses. The company offers a range of job-related courses which are open to qualified employees. Although machinists are prevented from programming because of jurisdictional agreements between Locals 3 and 8, under company policy they are eligible to attend NC programming classes. In a sense, their right of access to this information is sanctioned through company policy, while at the same time prohibited from its use by legal agreement.

On the floor, the machinists who use edits see it as a way to rectify errors in the program or to adjust for differences in machine performance. For them, this form of intervention is an extension of machinists' traditional skills. They discuss the use of edits as part of the process of 'getting the job done':

Your capable machinists on the floor use the edit to correct mistakes methods make. You talk to ****. I don't think there's a job down there that he got from methods that he ran their way. He made changes on it to suit the equipment they have. At the end of the shift, he changes it back to the way it was. In other words, the operator is forced to use his own ingenuity to get the job done, in spite of them. It's being done all the time.

I try to get involved in the NC machine, but you're really not supposed to. You're limited to what you can do. But I like to do the edits. The younger ones [methods] have a tendency to listen to us. If we see feeds and speeds that aren't right, or a better way to do the job – hold the job, put the job in a vice, a safer way – they'll make the changes for us. The older guys, they feel, 'That's how I programmed it, that's how you run it'. So, what choice do I have? I learned to edit by reading the book. I sit and read all the books. Whenever I get on a new machine, the first thing I do is find the book and read it.

There are more things to go wrong (with NC). There are a lot of unknowns and things you don't have control over. You don't know if a guy had a beer at lunch and miscalculated. So you take whatever steps are necessary to take care of mistakes.

The automation of planning the machining process poses new challenges to the division between conception and execution. Under conventional production, a machinist could ignore an incorrect OS and complete the process according to his or her own calculations without infringing on methods' jurisdictional rights. Because of the automation of planning, tape-controlled instructions simply cannot be disregarded; they must be physically altered for the machine to carry out the correct operations. This sets up disputes between machinists, who believe that their program adjustments continue to be a necessary part of skilled work and a continuation of a traditional practice, and methods, who view this intervention as a threat to their jobs and a breach of their jurisdiction.[7]

For machinists, editing (although contractually illegal at UFC) is a way of increasing and using their skills and remaining essential to automated production. It is also a challenge against the permanent removal of elements of planning, judgment and skill from the craft which continue to be vital to the production of accurate parts. The basic issue for these workers is: who controls the machine tools – the person who programs the tape or the machinist who operates the equipment? It is a fundamental question of shopfloor control.

Machinists struggle to continue to exercise discretion on NC, which is unquestioned when they work on conventional equipment, raises the issues of de-skilling and marginalization. More importantly, however, it indicates the extent to which the division of labour is both an historical and social process. Programming becomes defined as a technical rather than a conceptual task, severing it from the machining process. The difficulty with this division is that skilled machinists continue to be used as NC operators. In addition, they continue to work on conventional jobs, increasing their frustration with limited involvement on NC. Shaiken quotes a supervisor who is fully aware of the dissension this can cause:

A good conventional operator is a prima donna. You know that and you accept that. Back in the NC section, you are looking for very different qualities. The two don't mix.[8]

The removal of planning and the elimination of manual skill fragments the labour process and routinizes the physical operations machinists perform. The displacement of these skills cuts off vital information machinists have traditionally used to make adjustments in machining and the nature of control they exert in the labour process. Without these traditional forms of decision-making and control available to the machinist, the number of unknown variables in the machining process increases, creating greater risk of injury and higher scrap rates.

Machinists continue to depend on their remaining sensory skills as aids in monitoring the process. By increased reliance on visual and auditory skills they attempt to compensate for the loss of information acquired through manual control.

Not all machinists find this an easy adjustment. There are numerous complaints about their ability to apply the skills in quite the same way or with the same results. The machinists discuss the experience of sensory skill use on NC and the difficulties in its application:

> With NC, you're watching about three or four things move at the same time, so it increases the stress you're under. They have arrows and dials to let you know what's going on. You have the plastic glass that comes about this high (he indicates about shoulder height) and you have just about enough room to put your chin over it to watch it, and it becomes very difficult to work with. With the bigger NCs, they're completely enclosed in glass and you have to put your face right on top of the glass to see what's goin' on. There, the piece and the cutter are about five feet away. On those machines, you have to rely on the TV screens.

> On NC, it's very hard to listen to it – the machine makes so much noise. It's almost impossible to hear what's going on (in the machining process). You can't really listen to it – no way. All you really gotta go by is just by watching it and seein' what you end up with. Then you make your adjustments. On NC, I watch it. I watch everything I do. I watch very closely. I check every piece. Most people check every ten. It's more sight on NC than anything else.

> I use my hearing with NC. Things sound different because NC is running much faster. There's more noises involved, and a lot of those

noises have to be eliminated. That's not an easy thing to do, particularly at UFC because men are not on the equipment long enough to develop that sixth sense. We're talking about something that takes a minimum of six months to develop and to realize that you're using it as an aid. You have to refine your hearing [from conventional work to NC] and be more aware of it and if you work at it, it's not that difficult. But you have to work at it.

You also need to develop your timing. On conventional, you have a feel for the equipment. You know when a cutter's burnin' up. You feel the resistance. With NC, from an operator's standpoint, after you have the job set up and you're running it – well, that same type of rhythm is not involved.

Their experiences reveal distinct differences in the ways they use sensory skills on NC and the ability to apply them with confidence. The removal of manual work and control over planning translate into greater ambiguity for which the machinist needs to compensate. Sight and sound, therefore, become the primary, although less precise, sources of information.

Moreover, these skills are transformed from a central part of machining to a more marginal place, defined more in terms of crisis or intervention skills rather than an integral part of a practiced craft. It is a passive application of skill, yet the process demands a heightened visual and auditory awareness. As we have heard, the process of developing these skills has been uneven and difficult. In addition, some of the machinists have to rely more heavily on one sense, given the nature of the equipment, thereby limiting even further the extent of their control over the NC process. Others find the pace of automated work too fast to be able to accurately assess problems using limited sensory skills. For these machinists it exacerbates the loss of control over their work, creating conditions of uncertainty and tension.

Machinists also complain of boredom working on NC. The physical inactivity associated with monitoring and maintaining automated work engenders complacency while it continues to require a high degree of alertness and responsibility for the process. They talk about the frustrations of becoming marginal to the machining process:

When the [NC] machine's set up, I do a lot of daydreaming. That's one of the reasons I don't like it. It leaves you with nothing to do. It's boring – boring, so boring. You just follow the instructions they give you. On manual [machines], you can do it your own way. I'm thinking more – how I'm going to start the job, which technique is better.

You really get involved in the job when you do it manually. You want to see how the finished product is going to look. So, you keep pushing yourself. But now, sometimes when I'm finished setting the machine, I don't even worry about it. I just sit there . . . Boredom has a lot to do with automation. You stand up there, maybe you'll clean something. You do anything to keep yourself awake.

I see guys getting hurt more, working on the NC. Absent-mindedness is quite prevalent in there. If you don't have control over it, then you lose control over yourself.

Sometimes I just switch it into override so I can have a few minutes break from the monotony – it's monotonous. You really can't day-dream – even though there's nothing else to do because it's dangerous. You have to make sure the tool is in perfect condition. Sometimes the setup is loose. You can't take chances or play games or look away. You watch it the whole time the tool is moving. Plus, you know, it's so noisy in there.

To combat boredom, some of the machinists have fine-tuned their sense of hearing so that they are able to read or daydream while the machine is running. Lacking in the opportunity to direct the machining process, they develop this skill to the level of a 'sixth sense' in order to be able to focus their immediate attention elsewhere.

Today I got my first long-run job, almost three hundred pieces – a boring job. It's just a little piece with a bunch of holes in it – four minutes a piece. It's not enough time to lose yourself in thought. Every four minutes you gotta change the part. So, I open up the drawer on the bench next to me and I lay the magazine down on it. You try and read a magazine article on a four minute basis. Everytime you hear something different, you take a quick glance at it to make sure something didn't break or snap. Sometimes it's just a chip flappin' against something.

You're also keepin' an eye out for the boss, 'cause if he catches you, you're in trouble. I could, when I get to an interesting article that I would get carried away with, know just when I'm on the last tool. Without even realizing it, reach in the box and grab a part and just at the right time, walk over to the machine as the tool finishes, blow the thing out, take out the old part, put in the new one, and walk back and pick up on the exact word I left off on. Don't ask me how, but after a while it becomes second nature. It has nothin' to do with being a machinist.

Perhaps the most interesting perspective on the effects of NC on craft identity was given by a machinist who sat down a few days prior to our meeting and listed what he decided were the 'pros and cons of working with NC'. He said:

> Although I'm a trained NC machinist, I was surprised at the list of disadvantages I thought up . . . There's a definite lack of pride in the job, and it's diminishing every year. The job of handling NC equipment is perceived as a simple one. You tend not to get talked to with respect. Nobody (except for methods) will go over to the man at the machine and ask him anything about the job. They think that the machine and the guy who programs it are really the ones that do the work. So, there's a loss of pride.
>
> There's a big difference between ability and the job itself. You very rarely see a guy who'll say, 'I made that part'. Twenty years ago, there was a real pride in saying that. Now, a guy says, 'Yea, I made those. What about 'em? Are they messed up?' Nobody seems to care anymore how good you are. I do think it's a result of NC. Things that used to be extremely difficult to do well are now not even thought twice about on NC.

More than simply addressing the loss of craft or the elimination of skill, he frames this change in terms of the steady erosion of pride in the trade. This process becomes apparent, not only through the elimination of hand-practiced skills, but also through the increasing invisibility of the machinist on the floor. The stature of the machinist as problem-solver and artisan who earned respect and authority on the floor are no longer needed in quite the same form. The 'prima donna' of the shop retreats into a more mundane image of NC 'operator' whose primary activity involves program and machine maintenance. Planning and production formerly under their control are now divided and automated tasks, limiting their domain to certain phases of the operation.

It is interesting that the 'pros' of NC that he listed reflect more of managerial concerns for production, efficiency and lowering costs than a boost to machinists' skill or control over production:

> Well, NC can do things which is physically impossible on conventional. It does high quality work – better than regular conventional machines without buying very specialized equipment. To some extent, and this is debatable, the tooling can be a lot cheaper because you no longer need

special form cutters ground up. Also, some operators – if they work hard at it – can become good setup men without previous conventional background.

These experiences reflect the marginalization of craft skills with NC work. They confirm – in the self-perception of machinists and in the views of management – the extent to which NC is viewed as the replacement of highly skilled labour with automated processes. In addition, they support the evidence of a shift toward a greater role for and reliance upon the programmer in NC work.[9] The automation of physical craft work and the removal of significant aspects of decision-making from the floor diminish both the power and status of the machinist in the shop. Moreover, the inability to apply skills and experience to solve new production problems severely limits the ongoing development of the craft.[10]

The decomposition of skilled labour by further divisions between planning and execution and the mechanization of craft work is somewhat limited by the machinists' ability to circumvent this process. Clearly, however, NC has decimated much of the craft skills machinists once held.

However, it is critical to note that the trend toward the homogenization of skills is mediated by both machinists' resistance to this process and production needs.[11] Without the exercise of discretion, the production process would be completely chaotic. Not only would the scrap rate dramatically rise, but downtime for tape editing would be excessive. Machinists recognize their indispensability to the process and use it as leverage to obstruct attempts at further de-skilling and removing control off the floor.

COMPUTERIZED NUMERICAL CONTROL

Computerized numerical control, or CNC, is the most advanced automated machine tool the company has purchased. The machines are located in the general machining area, not in the NC room. The CNC equipment is referred to in the shop by its trade name, Hurco. Unlike NC machines which rely on tape-controlled, computer-programmed instructions, the Hurco's computer is based on the micro-processor and has a manual data input capacity (MDI). This permits direct operator input at the machine tool itself. It is capable of handling huge blocks of information for complex parts and, using its software, can communicate with the user to correct potential errors.

The Hurco is programmed through simple commands chosen from a menu format. No prior computer programming knowledge is necessary. It is part of the CAD/CAM system linking the design and manufacturing processes. However, UFC has chosen not to exercise this option, leaving design and manufacturing as technically autonomous processes. Compared with NC equipment, the Hurco is much smaller, quieter and designed primarily to machine small, but quite complex and often experimental parts.

For the machinists who work on them, the Hurco's capabilities pose new challenges in the application of trigonometry, metallurgy and tool usage to part production. It is distinguished from NC work by the ability of machinists to methodize a job right at the terminal using only a blueprint and the enormous range of options it affords in planning the production process. The versatility of the Hurco permits a much greater choice of operations which would previously have required a number of specialized machine tools to perform. The machinist is now faced with the possibility of conceptualizing the problem from a number of different approaches. According to Zuboff,[12] computer-mediated work translates conventional processes into abstract symbols. Exercising judgment is circumscribed by the logic of the system rather than simply by experiential knowledge. In the case of the Hurco, the system is based on a purely mathematical approach to and understanding of the machining process rather than a combination of sensory and mathematical skills.

Similar to NC machining, the Hurco requires the operator to exercise more conceptual skills in planning setups. In comparisons between the Hurco and conventional setups, workers use the same references to the removal of manual skill and the increased reliance upon conceptual tasks as NC work:

> Where you run into a problem is, for example, if you didn't make the initial spot big enough. The drill is working a little bit too hard to cut this piece of metal. It's gonna let you know. It's just not gonna sound right. You're not gonna get the correct kind of finish. So, then you make a correction in the program. You either run it slower, or make the spot bigger.
>
> Working manually at a drill press, however, you can simply press a little less hard. You hold it a little more steadily and press firmly, but slowly. All the compensation takes place right in your own mind and body. With the Hurco, it's isolated from the part. You can't feel the operation anymore. You've got to use your head a little bit more.

None of the machinists attribute de-skilling or a decline in control or craft to the Hurco as they do with NC. The complaints of frustration, stress and boredom common among NC machinists are not mentioned by machinists who work on the Hurco. In fact, they claim that working on the Hurco enhances their skills by giving them opportunity to control the entire planning process.

Having sole responsibility for organizing the work process recaptures for machinists those elements of the trade that have traditionally defined its fundamental skills. Conceptual work is the core of CNC machining. Control over this process situates the machinist once again at the centre of production. Moreover, the job of methodizing a part is integrated into the process, eliminating instructions from the methods department, relying exclusively upon the machinists' expertise.

Based on the blueprint and without the benefit of an OS or a tape, the machinist must decide the order of the machining process, the calculated feeds and speeds and the appropriate tooling. This is the same organization of work that characterizes conventional machining.

The similarities to conventional machining centre around the ability to exercise skill and judgment in planning the work. With CNC the physical skills are eliminated, as they are with NC machining, yet these machinists express no concern over the elimination of the craft work they described with such pride. The explanation for this contradiction can be found in their discussion of the use of traditional skills. The machinists who work on the Hurco are responsible for the entire process of complex calculations that convert conceptual and sensory skills into mathematical calculations and computer instructions. Basic machining skills remain as the core of CNC work – they are simply applied in a different format. As one machinist commented:

> I don't think there's a real difference between creativity on the Hurco and conventional machining. I suppose it's a difference in degree rather than in kind.

This notion of creative difference is a critical one. It is an indication that the nature of CNC work, as practiced at UFC, embodies the same basic skills as conventional machining. The geometry of the part dictates the complexity of the program. Using trigonometry, the machinist translates the path of the cutter into mathematical calculations.

Their descriptions of this process provide both a comparison with conventional and NC equipment and specific details of the nature of the skills and scope of control over CNC work. Programming the Hurco should not be confused with programming in the APT language. They are completely different processes. The Hurco uses a menu format which questions the operator on the specifics of the procedure, whereas APT is a computer language whose codes instruct the machine to carry out specific processes.

> Programming [on Hurco] is different [from NC] because there are no codes involved. It has a menu and asks you questions. Like multiple choice questions. They designed it for simplicity. But you have to be an experienced machinist in order to operate it.

> Taken in its entirety, I'd say the most enjoyable part of the Hurco is the setup and the programming. At least in this shop – 9 times out of 10 – you've never seen this part before. You have to apply the principles you know as a conventional machinist to make the part.

> I'm not restricted by other people's actions or lack of actions. I do my own programming. I set my own tools, make my own fixtures. Working on the Hurco is much more freedom of choice. Sometimes what you have running on the machine, the next job will – by changing the method of doing it slightly – be compatible with the setup you already have. The kind of freedom you have on Hurco is similar to working on conventional equipment.

> You're changing from putting pieces of masking tape down and markin' off your number readings which is what you do conventionally. Here you're putting information into a computer. You're using this same information in a different language in a different way. You're doin' no more, no less.

Programming or, more correctly, computer instruction, provides the bridge between conventional work and CNC machining. For the machinists, the opportunity to program eliminates the barrier between planning and execution and permits them to translate conventional knowledge into a series of commands and calculations.

The lines of demarcation between the machinists' and methods' occupations have been completely removed by virtue of the fact that programming has been designed into the system. Management's response to this technical take-over was to classify the Hurco as a conventional machine giving the machinist complete discretion over its use.

This decision can be understood in terms of the reorganization of work in the shop. The merger of job classifications eliminated all of the categories of specialized skills, resulting in essentially one general category of machinists. This allows management to use any machinist on the Hurco, regardless of seniority, thereby increasing the flexible use of skills at the point of production.

This reclassification has served a dual purpose for management. First, the simplification of the process and the elimination of the need for a computer programming language has enabled management to remove methods from any legal access to the machine's programming capabilities.[13] Second, the classification of the Hurco was part of a plan to merge all the classifications within the machinists' occupation, allowing management greater flexibility in the use of machinists' labour. They argued that the automation of manual work eliminated the need for those highly skilled manual machinists who operate specialized equipment. The flexibililty of the Hurco as a machine tool both in the kinds of complex parts it can manufacture and in the virtually unlimited variations available in planning a job is a challenge to the machinists' abilities to solve production problems. This process requires a high degree of abstract planning and an ability to apply the logic of machining principles in mathematical sequences. While the skills used in conventional and CNC machining are essentially the same, their application is based on different criteria. In conventional machining, machinists depend on sensory skills in order to make manual adjustments in the process. On the Hurco, these skills are converted into mathematical calculations and program instructions.

> If a speed or feed is wrong on a piece of conventional equipment, you can make an awful quick adjustment. But automatic equipment is gonna go exactly as programmed. You can't tell it, 'run about this fast'. You gotta give the machine specifics.

> A great many things that you do by feel on conventional machining, you have to do by numbers on a Hurco. For example, if you're drilling a hole. On a conventional machine, you've got your hand on the lever and you can feel the resistance of the drill. In Hurco, you don't feel the resistance. You have to tell the machine how fast to feed that drill into the material. It's simply a matter of translating from actual feel into numbers.

> If you need to know the length of an arc, you have to stop and figure that out for yourself before giving the machine a command. You'd have to have enough knowledge of geometry and trig to figure these things out. That's why some fellas like it so much, it can be very creative.

The importance of trigonometry in CNC work reflects the logic behind computerized machining. However, because the Hurco is based on a purely informational system, an indication of a good CNC machinist is his or her ability to transform experience, judgment and manual skills into a series of mathematical instructions. A machinist's ability to construct a sequence of operations depends on his or her understanding of the *logic* of the system and the ability to apply advanced principles of mathematics and the conceptualization of the entire process to this system. Without it, as one machinist said, 'basically all of the work will be "rule of thumb".'

All of the CNC machinists indicate that the most creative aspects of working on the Hurco involves problem-solving and designing programs. With the elimination of physical work, these become the primary expression of skill. Moreover, methods of operation are consolidated under machinists' control. Often, they will design and machine their own intricate and precision instruments and complicated patterns, challenging one another to plan more sophisticated approaches to their projects.

You can really use your imagination. I made a job about six months ago – a scalpel and some other surgical tools for my daughter who's a doctor.* They're gorgeous. I programmed them on the Hurco. It took a little while to do.

On the Hurco, your limit is your imagination. Now, take ****. He has a keychain so he doesn't lose his keys. So I say, 'We'll make a bell for it.' He says, 'Make it one piece.' I said, 'I don't think I can'. Now I'm startin' to think, 'How can I make a one piece bell?'

Machinists have figured out different approaches to programming the Hurco. Much of these variations depend on the intricacy of the job, potential safety problems and individual preferences for organizing work. One of the extraordinary advantages of the Hurco is the alternatives and options available in programming.

There are safeguards you can set up when you set up this machine. I usually set a slower rate of speed. In case something's goin' haywire, I have time to catch it. I don't have the reaction time some of this younger

---

* The specifics in this quote have been altered to protect the machinist's anonymity. The original piece had similar requirements for balance and manual dexterity.

generation has. When I setup it's at a considerable slower rate of speed than when I run.

Another aid I follow quite often is that you don't have to put the whole program for the whole part in at one time. You put in five or six blocks of information, check yourself out, go another five or six blocks and check it again until you get done. In other words, you don't have one hundred, one hundred and fifty block program in there. You can more or less isolate where your problems are.

I can do the same job three different ways, three different times and an entirely different way each time. I'm not restricted at all. I can look at a part and say, 'Which way are we goin'?' You can start at any of six different sides. Sometimes you just try one way against another. Sometimes I break it down into three setups if I don't feel good that day and just want to take it a little bit easier. You break it up so that you don't have so many checkpoints, so much to measure in each operation.

You can get pretty clever about the number of tools you have to use. I think it's clever, creative and satisfying to do as many operations with one tool as I can. There are programs built into the Hurco that allow you do things that you could never do on conventional. For example, to countersink a hole or a variety of different sized holes, you don't have to use a lot of different sized countersinks. You can use one small countersink and interpolate all the sizes. That's interesting to do.

The technical capabilities of the Hurco far exceed those that even a highly skilled conventional machinist can duplicate. Using this equipment, they are able to design programs for parts they have never been able to manually machine – even after thirty years in the trade. For these machinists, it is an opportunity to test their skills and learn new ways of applying them.

Working on the Hurco requires the machinist to focus on the mathematical properties of the job and to construct a set of instructions based on both mathematical calculations and a knowledge of tool usage. While machinists have always done this as part of traditional trade skills, this knowledge is applied in a completely different format. They describe this process:

You give the machine instructions to go from point A to point B to point C to point D. You go X up or Z down or whatever. Mill, drill, tape, bore, run around in circles. You can give it an almost unlimited amount of instructions.

Instead of milling, then later going on to the drilling – you're doing that all at one time – one shot. It holds two hundred and fifty blocks of information. The most I've used is one hundred and thirty and that was a phenomenal job. You tell it the length, width, how deep you want it and you give it one corner point. It has an information block where it gives all the tools.

The master block holds the basic information about the job – the diameter of the tool, its rpm, the calibration and the zero plane for the machine to relate to. The data blocks are broken down into different programming sections. Within each block are eight to ten different parameters which control the movement of the part and of the tool spindle.

In all of their discussions on programming, the machinists stress the critical importance of having a solid background in geometry and proper tool usage to be able to use the machine effectively. As one machinist comments:

Conventional and CNC involve, to a great extent, the same kind of thinking. You're still dealing with a problem in geometry and physics. But I think machinists' skills are going to have to develop and change a little bit. We're going to have to rely totally on the conceptual aspect of the job. We're going to have to know the concept of manufacturing – it's limited now just using the term 'machining'.

In addition to the application of traditional skills to a conceptual format, the machinist also needs to figure out different techniques for adjusting the machine's performance. An exact program may not be able to create a perfect part because of imperfect tooling or other conditions.

Sometimes you have to lie to the machine. For instance, you have a ½ inch end mill. You tell the machine 515ths so that it's cutting bigger, because the dimension is ½ inch, but the end mill is not exactly ½ inch. You put it the machine and you tell the machine it's 515. There's small things you can lie about. There are things that you know if the machine really did one way would not come out right, so you have to make adjustments. It's not in the book, you must create your own. The machine is smart, but you are smarter – you're a human. You have to think about how you're going to lie to the machine so he will obey you.

In a sense, this is a translation of the kinds of physical adjustments machinists make on conventional equipment when proper tooling is not available or a machine has idiosyncratic movements. However, the adjustments on the Hurco are based on the ability to find solutions through mathematical calculations and an understanding of the computer's logic system – an extremely sophisticated approach to solving machining problems. These adjustments are planned ahead and entered into the program to compensate for tool or other material imperfections. The exercise of independent judgment, while transformed from a sensory experience into a mathematical calculation, remains a critical element in CNC machining. Moreover, it represents a continued independence on the part of the machinist to apply his or her knowledge of the technical process to work around limitations. Interventions merely assume another form.

The only limits placed on programming are stored in the master tape and are designed to exclude instructions that are impossible to perform and/or are potentially damaging to the equipment. In instances where the program is incorrect, the machine will point out a discrepancy or missing information.

> If you put a cutter in that's too big, he's [i.e., the machine] going to tell you 'cutter too big'. If you don't put in the drilling cycle, the machine is going to stop and tell you, 'you forgot the drilling cycle'.

Machinists see this as an additional tool in programming rather than a usurpation of their role in the planning process. These machinists believe CNC represents a transition to skilled machining in the age of automation. According to their experiences, even though manual skills are lost, CNC work combines the remaining key elements of the trade – planning and organizing and controlling the work process, a solid understanding of advanced mathematics, tool usage and metallurgy with a powerful and flexible computer software system. The labour process is shifted from being an interaction of manual, sensory and conceptual skills to a translation and synthesis of these processes within a purely abstract form.

It is clear from these findings that the machinists' use of MDI is the critical factor in their ability to exercise these skills. Management's decision to classify the Hurco as a conventional machine, thereby assigning the occupation the task of programming, gave the machinists the primary role in CNC work. In some sense, the automation of manual processes homogenizes skilled machining. Those manual skills and the artistry of the craft are made redundant, eliminating distinctions between those who have mastered the skills and machinists who are less proficient.

Shifts in skill content embody contradictory tendencies. On the one hand, complex manual skills are eliminated along with the perceptual skills associated with performing these tasks. On the other hand, the mental work involved in programming demands more extensive planning and creative application of traditional skills, as well as the ability to translate the information sensory skills provide into programmable data.

Specific job hierarchies and occupational categories give form to the tasks and skills affected by automation. The trends exhibited by shifts in skill content take on particular organizational arrangements which are influenced by a number of factors: the existing division of labour and labour relations, the availability of trained personnel, the extent of automation, and the production needs of the firm. Job content is neither singularly nor primarily determined by automation. The reorganization of tasks and occupations assigned to carry them out is a managerial and, therefore, social decision. The Hurco's classification as a conventional machine tool – despite its being an automatic, micro-processor based machine – is an indication of the critical role of managerial strategy in creating different forms of work organization. This decision to reintegrate planning and, therefore, programming into the occupational category of machinists was carried out in spite of the fact that the task of programming is the work of another occupation – methods – according to existing collective bargaining agreements.

The difference between CNC and NC applications is instructive. Based on tape-controlled directions, NC requires a separate set of programming skills. This, according to the Office of Technology Assessment findings, 'allows programming to be separated from machine operations',[14] making it easier to be considered a technical process and assigning it to the programming department.

Since the job of providing operating instructions has historically, under union contract, belonged to the methods department at UFC, it would be a blatant breach of contract to attempt to re-assign this work to the shop if the company attempted to combine this technical work into a single job. What remained for machinists to do – setups and monitoring and maintaining the machine as programmed – drastically reduced the necessary judgment and time to proficiency to perform these tasks.

As with CNC work, there is the elimination of manual skills. However, the range of operations on NC within which to exercise judgment is considerably narrower. It is confined to elements of planning (setups) and solving problems which may emerge during the machining process. It should be remembered, however, that hiring policies at UFC did not change as a result of these lowered requirements. The company continued to advertise for highly skilled, *conventional* machinists.

The skills used in NC work vary according to the organization of work in the shop. Machinists in small specialty shops are often trained as NC programmers and operators (as several of those interviewed had been). The complex nature of the jobs and the limited resources of a firm often 'necessitates' combining these skills into one occupation.

The most widely reported use of NC documents not only the separation of planning from execution which NC allows, but also the opportunity to extend the division of labour on the shopfloor. There are the skilled setup machinists who are also responsible for solving most production problems, and the semi-skilled operators (or skilled machinists who are now used in this capacity) whose sole job is to monitor the process, push the Start and Stop buttons and load and unload parts. Most of the large, mass production corporations and many of the smaller sweatshops contracted by these corporations apply NC in this form.

SUMMARY

The effects of technology on skills are neither as clearcut nor absolute as is claimed by both supporters and critics of automation. The question of whether jobs are either re-skilled or de-skilled is too simplistic for the nature of changes underway in the automated workplace.

According to these data, machinists' skills are reconstituted with each form of automation, and which effects the different components of skill in complex ways. Undoubtedly, the manual skills that historically have been a source of craft identification as well as individual achievement and peer recognition in the trade are eliminated with the advent of automated machine tools. Nevertheless, other elements of machinists' skills – advanced mathematics, physics, metallurgy, machine operation and tool usage, as well as the remaining sensory skills along with judgment, discretion and experience – continue to be vital to NC and Hurco machine work, albeit used in different forms.

The data show the complex rearrangement of the components of these skills in different stages of automation. Without the opportunity to manually operate the equipment and apply those skills to that process, machinists rely on more abstract planning and problem-solving skills to compensate for the loss of physical control. The critical feature of NC is planning the machining process by programming instead of performing skilled manual operations. The highly technical skills associated with NC programming are fully separated from machining operations and become specialized skills assigned to the planning department. It is in this area that machinists have experienced the greatest loss of control.

The nature of the remaining planning work done on the shop floor requires that machinists reconceptualize the step-by-step procedures associated with conventional work within an increasingly abstract understanding of machining principles. Toward this end, NC setup is a more sophisticated application of this process than conventional work. Indeed, the threat of total decomposition of skills is mitigated by the critical and continued need for these skills. Nevertheless, the range of machinists' skills are utilized in a more limited context. Outside of the setups, application of their knowledge and skills is restricted to the monitoring and maintenance of the program. Although machinists continue to be a vital part of NC work, these are primarily *intervention* processes which leave machinists with a more passive role in the job. Machinists defy this marginalization, working out of classification by editing tapes at the machine. For them, such work is a fundamental issue of shopfloor control and an attempt to challenge the complete removal of planning from the machine shop.

Machinists' experiences with the Hurco provide them with a radically different role in automated machine work. This programmable machine tool builds on traditional machinists' skills using an additional tool (MDI) to produce parts. Like NC, conventional knowledge is applied to Hurco work in a more abstract form. The crucial difference here is that machinists program the job. There is no separation between planning and execution. Responsibility for programming eliminates the passivity and loss of control machinists are critical of when working on NC.

The reintegration of those processes associated with planning and the control machinists exert over the work is similar to the ways in which skills are used in conventional machining. The programming capabilities of the Hurco do not provide machinists with a new set of skills. Rather, it is a new application of traditional machining skills. Problem-solving and conceptual skills are more complex, however, since they are mediated through the logic of a computer-based information system instead of simply using experiential knowledge. It is within these areas that machinists reclaim those creative aspects of the trade traditionally associated with the use of manual skills.

The complex changes in skills which have taken place embody contradictory tendencies: de-skilling and upgrading; fragmentation and integration;

and shifts in the divisions between mental and manual work. As 'unfinished processes'[15] the forms they assume are historically and socially constructed. The evidence provided by these machinists indicate that negotiations over the use of skills and control over production processes take place at the point of production as much as through organizational decisions and collective bargaining.

5 Merger of Job Classifications

I don't feel the NC machines or any type of automation are going to make machinists into operators. If anybody is going to do that it will be the people who make the decisions, not the machines itself.

Machinist at UFC

The reorganization of work is taking place on the shopfloor and, where there are unions, in collective bargaining. Many of these changes, particularly in non-union environments, occur under the heading of 'job redesign'. Where a union exists, these job redesign programmes are in the form of contract concessions.

The central factor in both cases is to provide management with a more flexible workforce. The flexibility management has includes such changes as the elimination of job classifications and work rules, and the introduction of job rotation and work teams. The reorganization is often accompanied by a two-tiered wage system or another type of wage restructuring plan. In some cases, the contract is being replaced with a more 'flexible' agreement. As part of emerging labour relations policies, this process has been implemented in different forms and degrees.

The designation of the Hurco as a conventional machine also removes methods from programming its operations since, by definition, no programming exists for conventional work. Contractually, it gives management the right to train and use any machinist on the Hurco because it is classified as *conventional* work.

What emerges from this process is a restructuring of the organization of work and greater managerial control over the use of skills in production. However, this reorganization has different, as well as conflicting, consequences for the division of labour between machinists and methods. Contradictions over the 'ownership' of skills and labour processes by these two occupations become apparent with the introduction of the Hurco and its subsequent title as a conventional machine tool. The flexibility of computer-based automation removes some of the traditional constraints on the use of job skills. The historical division of labour which developed under Taylorism is now called into question both by the technical capabilities of computerization and managerial policy concerning the use of these capabilities.

Control over the labour process on the floor is not only affected by workers' ability to exercise skills in production, but also by the access they have to the acquisition of these skills. This is accomplished basically in two ways: on the job training and technical training through schools, courses and apprenticeship programmes.

The hiring and promotion policies at UFC for machinists are based on demonstrated skill as a conventional machinist. Job applicants and those considered for promotion are not tested for proficiency on either NC or CNC. Machinists are taught how to operate NC and CNC equipment through an on-the-job training programme.

Historically, machinists were shifted to the NC room, trained by shop leaders (more senior, experienced machinists) and left to work on the equipment indefinitely. Decisions on rotation and training were controlled by these leaders and were based on the union's seniority system. The machinists were training themselves not only to operate the equipment, but also to develop an understanding of the capabilities of the NC machine.

CNC has simplified this job, and the company uses machinists to methodize the work, with a less complicated format. Given the fact that there is less of a reliance upon highly-skilled manual work on these machines, the company has the capacity to employ machinists' skill with fewer restrictions. Several shop stewards remarked that, with the arrival of the Hurco, they saw the merger as a possibility three years before it was suggested by the company.

The purchase of the Hurco and other CNC sheet metal equipment brought with it a bonus for management.[1] Its advanced machining capabilities replaced the need for specialized skilled labour to operate that equipment. What management then presented to the union was a reorganization of the job classification system which would enable the foremen to use almost any worker – within some limitations of the seniority system – on conventional, NC, and CNC machines.

Classification for machinists – from the lowest to the highest rankings – are as follows: *B*, *A*, and *AA*. Specialized machinists include jig borers, ultra sound machinists and lappers. Tool and die makers are the highest classification in the shop and the most respected in the trade. What management proposed – and the union eventually accepted – was a merger of the three major categories and the elimination of the specialized titles, with the exception of the tool and die makers' classification.

As a managerial strategy, job reclassification allows the company to have greater control over staffing levels and promotions, unencumbered by traditional union work rules. As long as *B* and *A* machinists are required to operate nearly all of the equipment in the shop, there is diminished incentive for management to promote them to the *AA* classification.

Implicit in this job reclassification structure is the recognition that NC reduces the necessity for highly skilled machinists. The company does not extend the division of labour by using skilled setup machinists and semi-skilled operators, as do many mass production companies. Here, job reclassification provides a structure for cost-savings, in conjunction with the use of automated equipment. Flexibility in the production process is both the result of automated equipment which performs complex operations without highly specialized labour and the merger in job classifications which removes traditional union work rules.

The Hurco is also included in this reclassification process. According to the company, the Hurco is classified as a *conventional* machine and therefore, legally, any machinist in the shop is considered qualified to work on it.

> When it started out, an *A* machinist was trained to work on the Hurco. It was strictly an *A* job. The company always wanted it to be a *B* – strictly *B* – machine. That's what their intention was. Now, guys who are considered a *B* machinist here are really *A*, first class machinists. It's just a pay rate. It's how long you've been here. Then it (Hurco) got caught up in a union squabble (over the rights to program the machine), and then they compromised and said, 'OK, until we decide who's really gonna run it, we'll put *A*'s on it'. The union wanted a *AA* rating because you had to do everything. They wanted to give you more money for running that machine because you had to combine jobs. But the company had the attitude that, 'Well, we're going to train you and teach you how to work this. That should compensate you for the raise in pay.

This managerial strategy has consequences for the organization and control of the labour process – both bureaucratically and at the point of production. It has three main objectives: (1) cheapen labour costs by removing skilled workers from jobs which have experienced de-skilling; (2) lower these costs further by limiting promotions; and (3) removing control from the shopfloor by eliminating union restrictions on working out of classification, giving the company more flexibility in scheduling and production.

Several shop stewards saw this reorganization as a threat to the development of their trade, as well as an attack on shopfloor control.

When I went upstairs, they [management] were having meetings on it. I raised the question with **** – the boss up there. I asked him, 'Now, if you're gonna take all these *B*'s and put 'em into the NC room and leave them there, when they go for their *A* trial, they can't handle it. It's not really their fault. It's because you didn't have them on the machinery.' He said that if that happened, they'd just give him his *A*. Which I can't believe the company would do that. Besides, what's really important is that you can't really call him a skilled machinist.

I don't want to see the younger guys being stuck on NC. Other- wise, they're not gonna have their skills. When they go for their *A* trial, where do they put 'em? Back on the conventional machines doin' very hard jobs. Now, if they've been in the NC room for a couple of years they're not gonna have those skills, so how are they gonna pass their trial period? They're gonna lose their basic skills.

Under traditional union work rules, management might be forced to promote workers if they needed higher classification skills. Under the new agreement, the flexibility of the reclassification scheme gives management the right to disregard those traditional categories when assigning workers to jobs. Essentially, there are few restrictions placed on management. Lower waged machinists could be used on jobs that previously required a higher classification. Given the nature of the agreement reached between the company and the union, there is little incentive to promote machinists, regardless of their mastery of skills.

The fear expressed in the above comments that *B* machinists would become permanent NC operators with limited mobility and access to equipment outside of the NC room never really materialized. On the contrary, machinists have complained that training has been unorganized, too short and limited in scope.

On days, they only give them about a month on each NC machine. On nights, there's only a handful of us. The foreman I have on nights agrees with me – a machinist should be left on a machine until he feels capable of running it.

Until we had this agreement with UFC [job reclassification], the same guys were in the NC room and learned how to edit. Now, we're not really being trained at all. With the program we have now, you're not getting trained how to understand APT. They put you on a machine for six weeks – you're not gonna learn anything.

We really don't have a chance to learn what the machine is actually capable of doing. A K&T technician can make a better part in that (NC) machine than I can, because he knows the capabilities of the machine.

Essentially, machinists are expected to learn to apply their skills to setups and solving minor problems which do not require programming. Machinists are not encouraged to develop, nor are they given, necessary time and opportunity to understand the concepts behind NC machining. One machinist explained the difference between learning to be an NC operator and a skilled NC machinist.

Some manuals are written just to tell you how to operate the machine – 'page one, start the machine, page two, this is how to stop the machine'. Then there are other manuals that say, 'This machine has five axis machining capability, three simultaneous axes. It repeats within 1/10th, it is accurate to within 2/10ths'. That manual gives you the peripherals of the machine. It tells you the capabilities. What you need is that information and you have to learn the basic limits of the machine. And that's not taught at all at UFC. You're not supposed to need to know that. It's supposed to be the programmer's [methods'] problem. That's what really galls me about the whole concept of NC in the manufacturing field. It's totally misused. It's very deliberate. You gotta remember, that's exactly the way they [management] want it.

The training program is designed to limit machinists' control over the NC process and the NC room. However, the ignorance of the foremen about the NC process provides machinists with an opportunity to try and expand their sphere of control through experimentation and the application of NC skills they might have picked up from outside jobs or in classes on NC.

We have three or four bosses inside. They don't know how to run the NC room and they don't know how to run the Hurco. So what do they do inside? Nothing. But there's a *B* machinist and they give him *B* machinist money and the guy is running NC with methods help. And the Hurco, the *B* machinist is running that machine without methods or the bosses, because they don't know what's going on.

According to the reclassification scheme, the pool of machinists qualified to run the Hurco increases because the complex manual skills that would ordinarily be required for such complicated work are no longer necessary. A machinist, using 'rule of thumb' procedures on the Hurco would still turn out accurate parts. The elimination of manual work – and of the need for highly specialized skills – becomes justification for claiming a simplification of the labour process and, therefore, categorizing the Hurco as a conventional machine.

As we have seen, the machinists view the Hurco as a conventional machine because it reintegrates planning and control and requires a sophisticated use of conventional skills. The machinists' definition is based on the position that there is a *re*integration of traditional skills applied to an abstract, rather than physical, process. The exercise of control over the labour process which machinists claim as part of conventional skill is also a critical component of CNC work.

The notion that work is simplified, as management contends, is a distortion of the processes which are involved in automated machining. This position acknowledges that conventional skills are necessary in order to successfully operate the Hurco. However, it fails to address the form in which these skills are applied.

It is necessary to develop, as Zuboff states, 'theoretical insight and imagination'[2] in order to be able to fully utilize the technical capacity of the machine. According to the machinists, this constitutes an *advance* in the application of conventional skills, rather than a simplification of machining. On the one hand, the elimination of specialized and experienced skilled manual labour does constitute the simplification of the machining process: these manual skills are no longer necessary. On the other hand, the translation of these same skills into an abstract and mathematical format, which the machinist determines, adds another dimension to the machining process which also must be accounted for.

Simplification and reintegration are both outcomes of automation. They are, as stated in Chapter 1, concurrent processes – tendencies within the labour process. In this case, management has chosen to focus on the elimination of highly specialized manual skills to reclassify CNC work as basically a conventional process. It is a strategy that seeks to alter the terms by which the division and coordination of labour and, ultimately, the social relations at the point of production are determined. It is with this strategy that management approached the union on the issue of the merger of job reclassification.

According to the leadership and members of the local, UFC's management threatened to close the shop if the union did not agree to the merger. Pressured by this possible loss of jobs and a decimation of the local, the leadership pushed for a ratification of the agreement.

Essentially, the union agreed to the merger in exchange for a training programme which would rotate shop workers on NC and CNC equipment.

The agreement stressed the increased versatility of skills workers would gain over the present system of job classification that limited their access to this work, as well as a corresponding upgrading and pay increase for those in the lower classification. It was up to the membership to ratify this agreement. The members, however, were not so easily convinced of the benefits or of the threat of closure.

> The union was threatened that they [company] were gonna pull the machine shop. That can't be done as far as I'm concerned. They've threatened that to us so many times that, sooner or later, we're gonna have to put our foot down and say, 'Take it'. I don't think they could get away with it. They have too much money in there. They need some machinists there. For government contracts, they have to have a machine shop on hand.

> We had to take the vote three times until we got it right [he laughs]. We were very much against it in machining. The last vote was a tie. When we had the vote, everything was held back from the machinists. We didn't know anything until the last minute. They knew we weren't gonna like it. They [leadership] set it all up and he [president of the local] waited until the last minute. He had come down about six months earlier and got chased out of the shop. After that, he didn't come down until the week before. He handed us a piece of paper and told us what we were voting on. There was some discussion – really, it was a yelling match.

> It was voted against on the day shift and the night shift came in and the company sent down some muscle – some of their smooth talkers and they made a couple of promises to them and so it was a tie vote. Then it went to the executive board to be voted in.

Several issues in the merger divided the vote in the local: (1) control over the use of skills and the ability to exercise control on the floor; (2) the continued recognition and maintenance of specialized skill categories; (3) the inclusion of a seniority clause in the training and rotation program; and (4) assurances of continued and systematic promotions from *B* to *A* to *AA*.

Historically, unions have used collective bargaining agreements to define specific job categories and assignments, and rigid work rules to gain wage and job security.[3] This legal protection meant that unions could force management, through the grievance and, if necessary, arbitration procedures, to adhere to these specific contractual obligations. It protected workers and increased their power on the floor by (1) limiting management's ability to use workers in lower classifications for more skilled work at lower wage rates; (2) preserving craft control which can be used to limit output and exercise control over the labour process; and (3) giving skilled workers a respected position within the shop and a sense of accomplishment as specialists in the field.

There was discontent among the rank and file who saw these basic rights and protective measures coming under attack with the proposed merger.

> With the agreement – I have some difficulty with it because I wanted stronger language to protect each classification. As a person increases his knowledge, I want him to be paid for it. Basically, we didn't get language to restrict the lower classifications from doing a certain amount of sophisticated work. I guess we lost on that one.

> We didn't get a progression system for training. We don't really have a way of restricting the company from exploiting the *B* doing *A* work or *AA* work. If the supervisor on the floor wants to overly exploit somebody just because he knows he could get away with it, there's nothing we can do. We wanted training in seniority order which we didn't get. That was very important. You want your senior people to be more trained. It doesn't make sense if a *B* is more trained than an *A* or *AA* and is again exploited. Do I sound like Lenin? I keep using that word exploited, but I guess it's really the only way to describe it.

> The final effect (of the merger) was that there's no differentiation as far as classifications. Everybody is just garden variety machinists and, supposedly, different degrees of skill – *B*, *A*, *AA*. It takes away from your individuality.

> A person like **** who spent quite a few years learning to be a lapper, and the man has talent, believe me. He does work that nobody else there can do. ****. He was a methods man at one time. Now he's the lead man in jig boring. The guy is extremely gifted, talented. Now, he's just a garden variety machinist. It does take away from his individuality and the time spent in learning a particular trade. Now, they can just switch them around. People who have more specialized knowledge have no recognition any longer.

> It's strictly at management's whim who they put on what. And according to the contract now, with this new agreement, the union can't say a damn thing about it. Seniority doesn't play a part in it anymore.

They [management] can do anything they want. In other words, this agreement, going back to February 1984, gives the bastards a license to steal. And they use it . . . Hell, you can't file a grievance. I think the union was uninformed on this. I believe the union was sold a bill of goods.

With the elimination of specialized skills, management removed the power of these workers to withhold their labour through a work stoppage or slowdown. Giving management the right to train and work members out of seniority erodes the basic structure of the union contract, weakening the workers' ability to act in their own behalf on the floor and limiting their union's capacity to defend them through the grievance procedure.

As 'front-line' union representatives, shop stewards use the contract as a source of less formal, shop floor control. Armed with the contract, shop stewards enforce its provisions at the point of production. This power also gives them the leverage to permit temporary infractions of these work rules in exchange for some benefits for the affected workers. For example, a common practice is to allow some out of classification work in exchange for overtime or shift transfers. A few of the stewards interviewed often used this tactic and got overtime for workers in the lower classifications to increase their earnings. The merger severely limits their power to continue this practice.

They really got over on the union with this merger. Before, I wasn't allowed to go over and do welding. That wasn't my job, it was somebody else's. I was in sheet metal. They'd ask occasionally for guys who knew the job to be able to go over. But we would always negotiate something – upgradings, overtime. Overtime was a big one.

Let me give you a good example – the shear operators. We [shop stewards] were able to get a lot of overtime for them. Now, they're a labor grade 7. With the merger, they went up to a labor grade 11. He took a big cut in pay over the year, even with the upgrading because we couldn't get him the same overtime. Those guys wanted nothing to do with the merger. And you can't blame them. Their job security was a lot better.

We gave a lotta stuff away. When they wanted to work somebody out of occupation, they had to come and ask us and we would bargain with them. Now, it's not out of occupation. We [stewards] gave up a lot of power.

The merger severely limits and, in some cases, eliminates the capacity of shop stewards to enhance their members' earnings and protect less senior workers from performing higher-skilled jobs with no increase in pay, and weakens the ability of members to restrict output because of the lack of available labour. The removal of specialized classifications eliminates the power to withhold their labour since management is within their right to transfer any other qualified worker to do the same job.

In spite of this loss, however, shop stewards continue to use the agreement to enforce seniority rights and rules agreed to under the merger.

The company found they made some mistakes. They tried to get me off of nights, saying there's no more glass work. I said, 'You can't. I'm not a glass machinist anymore. I'm a AA and by seniority I can stay.' They made some mistakes and that's their problem. I'm makin' sure they hold up to their end of the bargain every which way there is. We use seniority bids to protect what we can.

The AAs (since the merger) are supposed to come back out on the floor to do the close tolerance stuff on the conventional machines. An AA will always be in the NC room, 'cause that's your leader. One of the little deals they made at night – only the AAs will prove a job out. That's one of *their* deals, and they're regretting they made it, because I made a stink about it a couple of times. I'm making sure they stick by it. They're tryin' to get away with it. I say: 'Uh, uh, not on my shift.' They want anybody to be able to prove out.

The data presented here suggests that both organizational and technological changes are responsible for shifts in the use of skills. Automation alone does not account for changes in the labour process. The allocation of work is not a technologically determined process. Its use is constrained by: (1) the specific needs of the company given the nature of its products and markets; (2) the particular skills and tasks needed in the production process; and (3) the cooperation and/or resistance of labour.

However, computer-based technology is highly adaptable to a diversity of production needs and provides a broad range of options and applications for the user. Within these parameters, the decision to reorganize the use of skills, and the particular form this reorganization assumes, is based on existing social relations.

HURCO – CONVENTIONAL MACHINE OR AUTOMATED MACHINE TOOL?

Process engineers face a different set of circumstances with the reclass-ification of the Hurco. Under the existing collective bargaining agreement, management is required to inform Local 8 of their intention to purchase new automated equipment and to discuss the role of affected occupations in the use of such equipment. Management violated this agreement when it did not consult the union regarding the purchase and use of the Hurco. According to management, notification was unnecessary since the machine would be operated within a conventional format. The argument over the capabilities of computerized machine tools as a reconstitution of conventional work or a new form of programming for process engineers became the battleground for control over the labour process.

This dispute prompted Local 8 to file a grievance against the use of the programming capability by the machinists. The nature of programming work – and its definition – were key issues in the arbitration. According to the union, 'the dispute between the parties concerns the continuation of the functions which Process Engineers have performed from time out of mind with regard to numerically controlled machine tools'.[4] However, management claims that NC programming is not applicable to the Hurco and, therefore, its use by machinists is not a violation of the contract. In the Letter of Intent to the union, management states:

> M.D.I. operations *parallel* other standard machine tools and in no way can replace N/C efficiency. *Instead of turning and reading dials, the machinist responds to visual statements and enters information using a keyboard that equates to operations and inputs that he would normally perform.*[5] (emphasis added)

Essentially, management argues that manual and sensory skills are simply transformed into keyboard responses to a series of software gen-erated questions. In addition, they raise the issue that the nature of the planning machinists are expected to do is not an infringement on process engineers' job jurisdiction. Management responds to the union's position that planning is methods' work by pointing out that machinists at UFC have historically been responsible for planning work on conventional equipment and often are expected to work without operation sheets.

Their argument on the nature and use of skills in CNC work raises two different, yet related, issues. The first centres around the definition of skills. Management contends that the nature of work and, therefore, of the skills required to do the job, do not change. In their stated position, they equate adjusting manual dials to achieve a specific cut with pushing buttons which instruct the machine to perform the same exact function. For management, it is simply a matter of different types of manual movements.

The second debate in the grievance focuses on the issue of programming. The company continues to support the right of process engineers to programme NC machines and their exclusive use of the APT programming language. However, management states that the conventional format which the Hurco uses does not require APT programming, or traditional operation sheets from the methods department. The machinist is capable of working directly from the blueprint. Therefore, programming is not a *feature* of CNC work.

The integration of programming and planning through the use of interactive software transforms the way skills are applied. This integration eliminates the need for specialized programming skills. It also re-establishes the link between conception and execution. Although the actual machining is fully automated, all of the decisions and planning necessary to produce the part remain intact. The historical cleavage which is fully developed with NC programming is virtually eliminated with MDI and with it the traditional programming functions of the Methods Department.

The union's position has been that *all* forms of programming fall within the process engineers' jurisdiction. Although specialized computer language is not applicable in this case, the Hurco does require special instructions using a computer terminal to communicate those instructions. In addition, the Hurco also contains a cassette tape feature for storing information. The union believes that the use of this feature is a clear violation of the contract since the production of taped information is the sole province of the Methods Department. When the tape is used, the Hurco becomes a numerical control machine operating off taped instructions. The union also argues the 'inefficiency' of machinist-run operations citing 'collective bargaining agreements and supplemental agreements between the parties requiring management to use its automated equipment in an efficient manner'.[6]

There are several components to their position. Like management, they offer a definition of programming. It is inclusive and focuses on basic programming *tasks* as opposed to the skills and processes involved. For example, preparing instructions for the machinist to enter data into the machine, and/or pushing buttons on the terminal to activate the system is viewed as comparable to producing computer-taped instructions. Under

different conditions (i.e., a shop with less skilled and experienced machinists) this position could be successfully argued. At UFC, the machinists
are highly skilled and extremely competent to operate such equipment.
The transformation of work to a primarily conceptual level reintegrates
their skills into the automated machining process.

Demanding the maintenance of the *task* of information input as well
as raising the issue of efficiency is extremely problematic. Implementing
this contractually would mean the reduction of a highly skilled and
experienced machinist to a simple operator – at the same pay rate – whose
only function would be to push the appropriate buttons on the terminal as
indicated on the program sheet and to monitor the machine. In effect, this
would constitute a duplicate operation, since MDI does not need computer
language to communicate instructions. MDI allows the machinist to bypass
those processes. To insist on the issue of efficiency as a means of retaining
control over the work actually backfires when placed in the context of
actual machining capabilities. The elimination of the need for specialized
programming skills removes process engineers from an undisputed place
in CNC work.

The union suggested a compromise in limiting machinists' work on
the Hurco to simple jobs – that is, the right to input up to eight blocks
of information. Data over eight blocks would automatically involve the
Methods Department. They cited precedent in an agreement made in
1967 over harness design.[7] Management countered that it 'was a real
dumb agreement' on their part and no such agreement would be reached
on the Hurco.

Behind this refusal, however, is the larger issue of managerial prerogative. Throughout the hearings, management continually referred to their
right to define the nature and use of equipment and, subsequently, who
operates it. According to management, since the Hurco is a new machine,
they were not ready to 'commit themselves' on its use. The question of
additional work rules to define automated work and job jurisdiction rights
is countered by management's insistence on increased flexibility in the
production process.

The grievance, taken to arbitration, was denied. The ruling concluded
that the Hurco 'did not transfer tasks from the Process Engineers to others;
it used a procedure that legitimately did not have to utilize customary Process Engineering activities'.[8] However, the ruling included an important
addition: 'questions remain open concerning impact on PE rights if and
when the Hurco is used with cassettes or with written material having
the predominant characteristics of an OS'.[9] In this case, the arbitrator
supported management's contention that the Hurco resembles conventional

equipment when the MDI capability is used. However, the ruling also clearly indicates that taping and storing jobs on the cassettes remain process engineering work – regardless of the method used to enter data on tape. According to the local's president, 'We lost the arbitration, but with the additional language on the use of cassettes, we actually won'.

For the local, the issues of skill transfer and efficiency of method as bases for the grievance are particularly difficult. As we have seen, the historical division of labour which encouraged centralized planning and specialized programming skills was never completely severed from machining. Planning remained an essential part of machinists' work, although a less formal process than methods'. The simplification of planning automated machine processes challenges the justification for maintaining specialized skills. Past practices *have* included machinists planning conventional work. That continues to be an integral part of their job at UFC. Although the *tasks* associated with planning are more easily defined, the delineation of the *processes* are unclear.

Computer software minimizes the importance of *specific tasks* in the labour process. Instead, it emphasizes the use of *information* to accomplish the same work. In the case of machining and methodizing, the information necessary to 'program' the Hurco is based on the same body of knowledge. The lines of demarcation which have separated occupations in the past were primarily established by task differentiation. The critical importance of information in an increasingly automated workplace transforms the specific tasks into a generalized application of knowledge about the entire process. Specific operations and/or tasks associated with the work are increasingly absorbed by the technology.

This transformation makes it extremely difficult to maintain traditional boundaries between occupations. The company argued that there was an exact transfer of skills from manual conventional work to the manual data input function on the Hurco. In fact, this is both inaccurate and misleading. While it is true that it is not traditional computer programming, operating the Hurco does require a comprehensive knowledge of machining and the ability to conceptualize the entire process. The step-by-step manual operations which include adjustments to the process while it is in progress are absent in Hurco machining. The operator must anticipate these changes and include them in the programming format. Management's superficial analogy obscures the exact nature of the skills machinists employ in the process.

The company realizes enormous cost savings through all these processes – the merger of job classifications, the elimination of methods from MDI input, and the categorization of the Hurco as a conventional machine tool. These changes are also an issue of managerial prerogative over the definition and use of skills. Management is no longer bound to work rules which once limited their discretion on the floor. These organizational changes reflect recent trends in collective bargaining agreements.

Renegotiating work rules assumes numerous forms – the elimination or reduction of job classifications, flexible staffing and sub-contracting, and 'pay for knowledge' schemes that increase workers' earnings as they master and perform more tasks. All of these processes significantly weaken the seniority system unions have historically fought and bargained for. Traditional work rules emerged with the rise of Taylorism. Management sought to limit the control workers had over the labour process through the creation of specific and narrow tasks which would be quickly and efficiently performed. The emergence of assembly line production methods solidified this division of labour.

At UFC, the merger of job classifications within the shop and the shifts in the lines of demarcation between methods and machinists reflect the process of developing flexibility in work arrangements and contemporary mechanisms for the control of production through the collective bargaining process.

6 Process Engineers

FREDERICK WINSLOW TAYLOR AND THE PLANNING DEPARTMENT

Manufacturing at UFC includes the shop (machinists, sheet metal and assemblers) and process engineering. Process engineers, or 'methods' as they are commonly referred to on the job and in this work, are responsible for preparing operating instructions for machinists to use in part production. They program NC lathes, milling machines and punch presses.[1] Planning is not an autonomous operation, but a supportive function to machinists' work. Although these occupations depend upon close cooperation with one another to produce parts, they are represented by different union locals. Machinists are represented by Local 3 and methods by Local 8.

Methods has its origins in the planning department proposed by Frederick W. Taylor. In Taylor's development of scientific management, the planning department was the centre of organizational operations for organizing and controlling all aspects of the production process. According to Taylor, its main functions were to include: '(1) order of work and route clerk, (2) instruction card clerk, (3) time and cost clerk, and (4) shop disciplinarian'.[2]

While most managers resisted Taylor's suggestion to centralize all of these processes within one department, many machine shops did establish a planning department to draw up specific and detailed methods for machining parts. The standardization of tools and the use of different kinds of metals for cutters and coolants all contributed to the development of specific metalworking procedures.[3]

In his attempt to systematize machine work, Taylor stressed the importance of developing precise and scientific methods for determining correct feeds and speeds and appropriate cutting tools and metals for efficient production. Essentially, this meant subjecting all of the information machinists use in their work to specific mathematical formulae and transferring the responsibility for this work to the planning department. Methods, as an occupation, was conceived and developed solely through the separation of planning from the execution of machine work.

Taylor had developed elaborate slide rules and detailed charts for calculating the capacity of different tools to cut various kinds of metals. Although

machinists also use these slide rules, many of them were, in fact, designed
for the planning department as their 'tools of the trade'. It was possible,
as Taylor pointed out, to create a specialized occupation of workers who
had no machining knowledge or experience to plan the process. Over the
course of a twenty-six year experiment, Taylor observed and recorded all
of the machinists' actions, measuring each process in terms of its efficiency
and accuracy and develop precise (and efficient) production methods.
His attempts to remove all of the discretionary elements from planning
ultimately failed. However, the codification of the twelve variables involved
in machining and the in-depth analysis of its scientific, metallurgic and
mathematical properties enabled him to systematize much of the planning
process. In effect, methods – who generally have little or no machining
background – perform the same conceptual work as machinists, relying
on a strictly scientific and mathematical approach to metalworking. For
machinists, planning is based on both technical training and experience. A
change in the social, not technical, division of labour which gave rise to
a separate planning department.[4] Analyzing the history of this occupation
and the processes that became the province of methods is important if
the present changes in advanced automated machining underway and their
impact on methods' work are to be understood.

The work of methodizing or planning a job is carried out in three different
ways: (1) calculating all the coordinates for the job manually and preparing
an operation sheet (OS) which describes the step-by-step procedures the
machinist should use, including the type of manual machine to perform
the operation; (2) calculating all of the coordinates and, using the APT
language, programming a tape for the NC machines; and (3) using the
CAD/CAM system, which contains a software package that performs many
of the complex calculations for some of the more intricate parts, creating a
tape for NC work.

Until almost a decade ago, the methods department at UFC was allowed
to recruit top notch machinists from the floor. An in-depth knowledge of
machining and some familiarity with programming from their work on
NC, was sufficient for promotion.[5] A basic requirement of the job is
a solid background in advanced mathematics – calculus, geometry and
trigonometry. Those hired from outside the firm are required to have an
AA degree in mechanical technology or related discipline, and a minimum
of five years' experience in the field.[6]

More recently, the promotional ladders into methods from the shop
have been blocked by these degree requirements, as have been those
opportunities to advance from methods into manufacturing engineering, which now requires a four-year degree. However, the demand for

educational qualifications does not necessarily reflect increasingly complex or advanced work, nor is there strict adherence to this policy. As one process engineer stated:

> What that has done is to create dead end jobs for a lot of people. It has taken a lot of incentive out of the jobs . . . But they did promote three guys last year who didn't have degree requirements. And the only reason I feel they were brought up is because they needed some people who could come into the department and start doing work, since there was a high work load and some people had just retired. They needed people who were able to come in and start working with a minimum amount of training.

The relationship between planning and machining is made clear as he continues:

> The people who were picked up were . . . trying to get into the department. They made it their business to learn part of our job . . . They, of course, knew the machining end of it so they were able to come into the department and with a little bit of training in programming and learning the different sub-routines they were able to start producing. And that was a lot faster then people who had Associates degrees from the outside.

His observation debunks the myth of the importance of degree requirements to perform the work. In addition, it points to the extent to which machinists are able to do methods' job because of their expertise in machining. A few of the more senior workers in the methods department were promoted from the machine shop and/or had previous methods experience in another company. Most of those with less seniority had little or no direct experience with machining. They received training primarily in state and community colleges through programmes that focus on basic manufacturing applications.

Essentially, methods is responsible for coordinating all operating instructions for the machine shop, and making decisions regarding special finishes from sub-contractors, and is an intermediary between occupations involved in the different stages of design and production. As one process engineer said, 'We have to take a part from raw stock to finished stock. We are the ones who write it all down'. Another commented in more detail on the pivotal role they play between design and manufacture:

We are sort of a go-between for the manufacturing engineer, the machinists and production control. An engineer, be it chemical or mechanical, is the one who designs the product. The manufacturing engineer works it out with them to see if it is feasible to manufacture. He's the one – after the blueprints are drawn up – who comes to us to manufacture the part. A lot of times we have to go back to them and even back to the engineer himself and tell them, 'No, you guys can't manufacture it the way you want to. We'll have to make a few changes'.

Methods' work is defined in terms of the feasibility of the manufacturing process – the ability to produce a part as cheaply, efficiently and accurately as possible, with an understanding of the capacity of the machinery, tool usage and safety. Methods must understand both the needs of engineering for the use of the part and the ability of the machining department to actually produce it. Several describe their role in the production process:

Basically, it's not the same thing day in and day out. You are always working on a different kind of part. You always get interrupted. I always have to run downstairs because they're running one of my jobs and I have to correct something or interpret something for them.

I have to work with the engineer if he wants to build something. I might have to tell him, 'Hey, we can't build it that way. It's impossible. You can't turn it inside out'. It entails a lot of different things. You are always on the go . . .

UFC works out a system where we . . . rotate in the shop . . . for about two or three months. But if it's a job of yours and it has your name on it and there are problems with it, you usually will get called down. Unless it's something very simple that the shop methods man on the floor can see right away. But if it's on the automatic NC machines, you might have been working on that job for a month, he [methods person assigned to the floor] can't figure it out in ten minutes. So he has to call the man who actually planned the job and he would have to come down [on the floor].

Methods' close collaboration with the shop in creating an accurate program and OS is evidenced in these comments:

We go down and talk to some of the guys [in the machine shop] we have gotten to know over the years as having a good head on their shoulders. We'll talk to them about specific problems and how they would handle them or how they would like us to write up the OS. So we interact with them a lot . . . Sometimes we make blatant errors and sometimes we make minor errors and they want to discuss it.

The standard practice in the field is ruin the first piece but not the second. The only time you don't ruin the first piece is if it's very expensive material. So we go through that stage with them [machinists] and work with the machines. He points out things he doesn't like about the program and we sometimes change them or possibly we would do them his way in the future.

These descriptions point to the interdependence of the two occupations and the critical importance of consultation with NC machinists on appropriate machining methods. Without this kind of collaboration, methods would have enormous difficulty creating accurate programs. All of the work done in the shop by machinists, sheet metal and assembly is methodized by process engineers. Methodizing a job into a series of steps is done by machinists as well, except that it is primarily a mental rather than a written process. Every machinist performs these processes in preparation for a job as part of the setup. As one process engineer commented:

Although back then [as a machinist] I didn't write it down on paper, it was the same kind of procedure. Now, all we [methods] do is write it down on paper and somebody else does the [physical] work.

Methods' work is divided into two parts: the actual planning and methodizing of a job, and trouble-shooting between engineering design in its conceptual stages and production to de-bug the NC program. As we have seen earlier, smaller shops often depend upon machinists to methodize their work and consult with engineers on the feasibility of manufacturing the project. At UFC, however, the size of the firm and its output of highly complex parts would make it extremely time-consuming for machinists to also methodize all of their jobs.

A process engineer provides a description of the steps involved in methodizing a job:

The prints come down from the design department. From that point, I would study the print and if it's assembly – separate it into details.

From there, I will work on parts individually. I will decide whether the part is to be punched out with a punch press or machined . . . If it can't be manufactured, I would tell them it had to be purchased. From that point, I make a decision whether I want to make the part on a CAD/CAM system or if I want to write it manually and produce it with an APT tooling program or if I want to produce it by the machine language method, which would be a direct method of obtaining machine language right onto the tape.

He goes on to explain the reasons for choosing one form of technology over another to methodize a job:

The complexity of the part, the type of operation [determines the choice of technology]. A complicated milling job – let's say a job that has a lots of pocket areas, which is like hollow cavities in it are given to a CAD/CAM system because there's a lot of sub-routines already in the software of the CAD/CAM system. If it was a part that was more simple, I might want to go to the APT language. If it was just hole drilling and some basic simple paths in milling, I would probably also go the APT route. If it's a very simple part, let's say with five or six holes and bored sections, I would write a manual operation.

What distinguishes planners in the methods department at UFC from those at other corporations is that they remain all-around process engineers involved in every aspect of manufacture and assembly. One process engineer discussed the organization of work at UFC:

The particular work we do is challenging – it's interesting. It could get boring if it was a plant that made one type of thing all the time, or we just specialized in one type of product where the design didn't change too much. At UFC, we do so many different kinds of parts it makes it more interesting. I know some companies are structured such that they [methods] specialize in one area. They may only do machining or sheet metal or tool and die making – just one particular area. We do all that . . .

With each different machine you need different skills. We do have to know a lot more (than those who specialize). There's no question about that. Some companies even have it designed where they have a programmer who just programs and the man who writes the OS's. A tool designer would only design tools. We do all that . . .

If you're in charge of the whole ballgame on that particular assembly or detail, there's less of a chance for mistakes. Except that it can only work up to a point. Take **** [a local aircraft manufacturer]. One man couldn't do tool design *and* program. You have parts as big as this house. There's no way he could do all of it. So he *has* to specialize.

Another described the characteristics of the job in terms of the diversity of tasks:

Methods men wear multi-hats. We have different jobs to do – there are different parts to the job. Sometimes it's a go-between, sometimes it's writing up the OS's, sometimes it's arguing with machinists. We do all kinds of things. It's where you more or less have to know a little bit about everything in order to get your point across – in order to get things done.

UFC's production needs and the history of its labour negotiations have created a highly diversified occupation, expected to work in every phase of manufacture and assembly planning. Given the lack of specialization and differentiation, methods' work will be analyzed here in terms of the changes in the nature of the skills needed to perform the range of duties the job entails. Like the machinists, methods are highly skilled and versatile. The range of tasks and the knowledge of different aspects of manufacture and assembly to do the job require a broad understanding of both design and machining. Their responsibility is to translate the dimensions of a design into a series of machine operations onto an operation sheet or NC tape.

As a 'service' occupation, the introduction of sophisticated technology poses a unique problem for methods. As design work and manufacturing are increasingly automated, methods is also affected, as in the case of the Hurco which essentially eliminates their programming work. At the same time, methods also has experienced changes in the labour process with the introduction of CAD/CAM. Both the direct and indirect effects of automation will be examined here in order to understand the nature of these changes.

Methods design procedures for machinists to follow which define the closeness of dimensions and tolerances of the part. Conventional jobs are planned by choosing a particular machine (for example, a lathe or milling machine) and a series of steps to be performed which will facilitate the machining process. Methods must know the capabilities of the different machines, the specific operations to be performed and the order in which they should be done to achieve the needed tolerances and dimensions. Methods are trained according to the same technical criteria as machinists regarding metallurgy, machine use, and mathematical applications. However, since they do not actually machine the parts, greater emphasis is placed on knowing the technical capacity of the equipment and tools, and the principles of advanced mathematics.

Numerical Control Planning

NC planning extends the conventional planning process with advanced multi-purpose machines and sophisticated programming techniques. NC machines are faster, more accurate and versatile than conventional equipment. The programming system, APT, was developed to design elaborate and complex parts and is capable of a wide range of applications. As Shaiken points out, 'in its full-blown version it is powerful enough to program virtually any surface that is mathematically definable'.[7] One process engineer explained his satisfaction with working with NC programming this way:

> You're taking something from a mathematical standpoint, which is in the Cartesian quarter plane – X+Y+Z – and all these different values turning into electronic impulse by tape and then have it [machine] move within tenths of a dimension going to that location.

Unlike machinists who discuss their identity as craft workers in terms of manual skills, this description focuses on the use of math and conceptual skills to instruct automatic machine operations. The integration of these different processes – a conceptual understanding of the process, the mathematical calculations of the dimensions and tolerances, and its translation into a set of programmed instructions – constitute a definition of methodizing part production.

CNC and NC tape machines are essentially the same machines. They both are programmed in APT directly or with a program generated from the CAD/CAM system. The major difference is that CNC machines have a mini-computer attached to the machine. This allows editing to be done directly at the terminal rather than by removing a tape and editing it in the methods department, as with an NC machine. The computer also allows the tape to run from memory. This is particularly useful in long running jobs. The tape is made of Mylar and sometimes breaks or tears, the holes get frail, and dust particles can collect on the tape, misreading the program. For methods, this means spending less time editing tapes in the department and easier access to edit on the floor. Machinists who have learned APT also have access to editing, although this work remains out of classification. Some machines run strictly on tape (the older models), others are programmed onto cassettes which are loaded in the machine; the newer machines can run strictly off memory.

The machines also vary in capacity, as a process engineer explained:

> On the K&T (Kearney and Trecker), we put on bigger, more complicated pieces on there because they have more tool capacity and they are more rigid. We've got K&T200, K&TVB4, K&T180. Then we also have some Pozi-Tools, a few Hurco's and the Travelon which is a lot bigger. We can put a piece on there forty inches long by twenty inches wide and do the whole piece with no problem. You couldn't do that with the Pozi-Tool, but the K&T would be able to handle it.

Each machine requires methods to learn some new language although they are all essentially the same. However, it has increased the versatility of the programming skills methods already possess. One process engineer compared the additional information he needed to program with the changes the machinists face:

> I think it's a lot easier to learn a language change. And besides, if we put in the wrong language and we process it, it will come back and tell us – 'Hey dummy, you did it again'. So we have safeguards in there.
>
> With the machinist, he will have to learn new controls because although they *do* the same thing, they are completely different. So he has to learn new controls, new ways of setting up a job, new ways of holding the job. I think it's a little more complicated for him than it is for us.

Although the machining process remains the same as with conventional work, the versatility, complexity and variety of NC offer methods a challenge. The power of the APT language makes it a very useful tool in planning such complicated processes. But, as Roberts and Prentice[8] point out, the standardization of tapes and control commands provide a means of simplifying these variations. They provide a simple explanation of how APT works: 'The numbers and letters punched onto the tape can tell the NC machine how far to move, in which direction to move, and

in many machines, how fast to do it and which cutting tool it is to use.'[9]
According to a process engineer, basic APT programming consists of:

> . . . look[ing] at the print and defin[ing] the part in terms of geometry
> – surfaces, holes, points, planes, which are depths of cut or drill depths,
> milldepths, depths of a hollow pocket. Define all the geometry in 3D
> space. Then you would set up the motion parameters – you plan: How
> do you want to machine it? With all the surfaces defined, we'll run the
> tool around the surfaces of the part and that would cut the part. There's
> all the curvatures and shapes you have to consider – a flat top on it –
> if you mill flat, you run the cutter along a series of lines along the top.

Basically, he describes the same conceptual problem-solving process
machinists use in creating their own setups. However, methods – unlike
the machinist who also uses sensory skills to plan a job – must rely mainly
on the knowledge of machine and tool capability in order to create an
accurate program. It is at this point that collaboration with the shopfloor
is vital. As we have seen, machinists on conventional equipment would
simply make a manual adjustment during the process to compensate for a
problem with a cutting tool or a difficult procedure. Automated machining
eliminates this kind of intervention. Therefore, methods needs to have a
thorough understanding of APT in order to deal with uncertainties that
arise in the production process.

Programming defines the identity of the process engineer much in the
same way that sensory skills to a great extent define the machinist's
craft. The ability to produce a part, whether through manual operations
or programmed machine instructions, is the source of 'ownership' of the
labour process. Methods claim NC work because they create tapes which
instruct the machines to perform a series of operations. In a discussion
of his role in machining, one process engineer describes this notion of
ownership:

> On NC where I have to make tapes, I will not only tell the man what
> tools to setup, I will make a tape for him which he will put into the
> machine. All he has to do is setup the tool and the vise, whatever it
> might be to hold the part itself. *He holds the blanks and I make a new
> piece, not him.* (emphasis added) *I* put in the tolerances – he might have
> to set the depths for the cutters, but *I* make the whole part.
>
>   On conventional equipment, he will actually make the piece . . .
> Whereas in NC, he has no choice. He sets up the tools the way I
> tell him and he uses the machine that I tell him and he uses my

tape . . . He has no say in what kind of end mill or drill to use . . . There I am specific – I want a one inch end mill for a reason; I want a quarter inch drill for a reason. So he has no choice in the matter.

Another process engineer appropriated the machining process in his statement:

We run NC machine tools which are controlled by tapes which we produce in our department.

Job jurisdiction rights place programming under methods' control. However, methods' claim to the tape-produced parts indicates the extent to which control over a process constitutes ownership and the subsequent marginalization of machinists to that process. There is no corresponding sense of control over written instructions, for conventional operation sheets can be easily ignored by machinists producing the part. Programming automated machine tools provides methods with a critical role in machining. Planning a job through tape programming gives them nearly complete control over the actual machining process. The shift in production (from conventional to NC) technologically re-defines and transfers control to the programmer (in this case, methods).

COMPUTER-AIDED DESIGN/COMPUTER-AIDED MANUFACTURE

Advances in design and programming technology have had an impact on methods. CAD/CAM, a design and manufacturing computer system, is used to generate programs for C/NC machines. Unlike CAD/CAM systems in mass production facilities which produce direct, machine-readable programs, the CAD/CAM system at UFC is not directly linked to the machine shop. Therefore, the spectre of CAD/CAM systems automatically producing parts from programs generated in the methods or programming departments (or, more futuristically, from an engineer's computer terminal) is non-existent at the plant. However, methods use the software to develop NC tapes. Working with the APT system, methods must perform extensive mathematical calculations to arrive at accurate machine instructions. The software in CAD/CAM, however, eliminates the need for such extensive and detailed work. Every process engineer commented on the speed and accuracy of the system's ability to perform this function. In addition, they raised the point that since both parts and machining capabilities have become increasingly complex, manually performing all of the calculations would be extremely time-consuming.

Methods develop their own set of programs called macros. A macro is a sub-routine, a set of programming instructions and sequences which is generated through CAD/CAM software. Methods design them to deal with certain complex machining procedures which can be applied to a number of different jobs with similar complex sequences. The ability to generate a library of these sub-routines indicates methods' proficiency with CAD/CAM software. These canned programs eliminate the need for trigonometric work for defining tolerances and dimensions. However, these instructions are limited to specific kinds of parts. One process engineer provides an example of their use:

> Macros are a little step in the system . . . But they can't always be used, even when you're making a similar part. It may be designed only for a square or rectangular part. There are always these variables that come in.

Although nearly all of the process engineers found macros to be extremely helpful, a few thought they regimented the process. One process engineer described the steps involved in the CAD/CAM process:

> I go down to the CAD/CAM room and think about how I'm going to approach this. Then I draw the part, but not in detail (as drafters do). I do this on the screen by calling out lines and curves and circles and geometric figures on the screen by using a light pen or typing it in.
>
> I can call out the coordinates of a line and the line will appear on the screen. I can call out the center of a circle – that would give me the radius or any other information and I can have the circle on the screen. That's not the complicated part – that's the easy and fun part.
>
> After that I get a cutter motion [path of the tool]. I put a little circle on the screen and that's my cutter. Now, there's a few ways of doing that. One is to use one of the existing programs in the system. I just call out one line and I will get the cutter path. Two, doing it manually which is complicated, or maybe the sub-routine in the system is not suitable for the job I'm doing. Then I would either create my own macro or do it manually where I would walk the cutter around the geometric figures I want to go to.

Once I have figured out all my coordinates for the path of the cutter, I will take this information and punch it on an ASCII tape, and send it through a mainframe computer, which translates it from an ASCII tape to machine language so that the machine can understand it. Then I write out an operation sheet. It explains what other operations might have to be performed before the NC machine applies its little magic.

CAD/CAM work is basically done in two stages: the first involves defining the basic parameters of the job; the second is the creation of a program to instruct the machine on the correct sequences of operations. It is primarily the second stage that demands applications of more sophisticated machining principles. CAD/CAM programming is designed to facilitate the extensive machining capabilities of automatic machine tools and the production of intricate parts. The complexity of the sub-routines in the system, as well as those which methods design, have contradictory effects on methods' work. On the one hand, it simplifies and performs, in some cases, the enormous amount of mathematical work which more complicated machining processes require. On the other hand, it demands a greater conceptual understanding of machining principles and the ability to apply this knowledge to programming.

These contradictions have been crucial in salary negotiations. Management has focused primarily on the simplification of computation skills and the use of advanced mathematics to justify their denial of pay increases. As one process engineer discussed:

One time an executive board member tried to get us an increase in pay because we require a new skill to program these machines and to run the CAD/CAM equipment. The answer that he got was that we should be getting paid *less* since the programming makes it easier for us. We don't even have to know math to program because the machine will do it for you.

Methods stress the infinite ways of doing one job on CAD/CAM, since the flexibility of the system allows them to try different problem-solving approaches. Like machinists, methods find this a creative challenge – the possibility of using their knowledge in new and different forms. One process engineer discusses the way he approaches a job. His description is very similar to a machinist working on the Hurco:

First you check the print and then you begin to plan the job. You may sit there for ten to fifteen minutes just thinking ahead – 'How am I going to plan it?' 'What should I do first?' 'How should I set it up?' 'What machine should I put it on?' 'What type of technique should I use to obtain the automatic tapes for the machines?' It may even take a half hour just sitting there thinking what you should do to it – the most efficient, easiest and accurate way. You're doing a lot more different areas combined. With conventional planning, it would be a step by step process – broken down into different parts.

Programs in the CAD/CAM system reflect a more integrated machining process wherein a number of procedures are performed simultaneously. Methods instruction reflect this approach. Although APT also programs these same machines, there is a significant difference between them. Process engineers explained it this way:

> When you go through APT, which is a computer language, you have to know the *language* – you have to understand how it [language] works. It's a very good computer language. It's probably the best designed language for machining . . . But when you compare it to CAD/CAM, you require another type of skill. You may not require the programming knowledge, but now you're required to have a different kind of knowledge. For example, 'What *information* do I really need to produce this NC as an end result?' 'How can I do it the fastest and easiest way and the most accurate way?' . . . You don't really define the same things as you would in APT on the CAD/CAM system. Similarly, but not exactly the same. Because of your interactive use with this computer and its sub-routines already in the software – you can call up and use it.

> Now, in the past, you would have to write the geometry in a computer program by hand and calculate it out and then you would have to write your cutter path around it. Depending on how difficult the geometry is, it could be difficult to write. The computer, putting in the geometry, even if it's quite complicated, just takes a couple of minutes. Once it's there you just say, 'I want this contour to be machined' and describe the tool and it does it.

The key difference here is the way in which methods' knowledge of manufacturing principles and technology are used. CAD/CAM is primarily an information system. Unlike APT, which is a specific language methods apply to develop a set of instructions, information is presented to the CAD/CAM user in the form of software packages. In order for it to be used effectively, a process engineer synthesizes all of his/her knowledge about manufacturing through the software. The approach to problem-solving is guided by the logic of the system. How skillfully methods can use the software depends on their ability to communicate their knowledge to the system. As one process engineer commented:

> The computer gives you more choices as to the approach you would use . . . You can literally reproduce any shape onto a piece of metal. Years ago, that would have required quite a bit of skill and ingenuity, not to mention the time involved . . . Because of the computer that is now possible, because you can actually take the measurements and pass them onto the computer and you will get the coordinates to reproduce that shape.

Zuboff notes that theoretical insight and imagination are vital in computer-mediated work.

> By creating a medium of work where imagination instead of experi-ence-based judgment is important, information technology challenges old procedures. Judging a given task in the light of experience thus becomes less important than imagining how the task can be reorgan-ized . . . based on new technical capabilities.[10]

This does not automatically imply a loss of skill or that less skill is required to perform the tasks. In the case of process engineers, the management and use of information is transformed into a more abstract process. However, it does not diminish the theoretical and technical knowledge demanded of process engineers. What might have sometimes been an 'educated guess' on feeds, speeds, tolerances and/or procedures – partly an experiential decision – is now a more exclusively scientific-based process. It appears to be a more focused approach to methodizing. One process engineer describes his experience with CAD/CAM:

> You're now able to produce more greater accuracies with less steps and plus with the power of the software you are eliminating a lot of steps and effort. But, you've got to know another whole world . . . Now you're dealing with a sub-routine that's in the software and you are putting in variables in terms of the feeds and speeds and it calculates for you. But you still have to check them and you have to have more of a knowledge (a vast knowledge) of data on feed and speed knowledge . . . The methods man of today has to know a lot more theory about actual cutting

science . . . It's more scientifically done because of changes in the nature of the equipment as well as the equipment that he's working on himself – CAD/CAM.

Other process engineers provide these insights into working with CAD/CAM:

> It does take quite a while for you to actually grasp it and see all the things that it can do for you. It's basically not a system that you can use by itself, but an addition to what you already do.

> You have to know your field and then you go on to learn the logic of the system – the operations in the computer. All the skills you already know, including blueprint reading – trig and calculus all come together and you apply it to CAD/CAM.

Instead of generating all of the data needed to methodize a job, they extract much of it from the system using the software and commands. However, basic knowledge of manufacturing technology and processes continue to inform the work. Methods' skills in part manufacturing are influenced by the availability of software as well as the complexity of the parts. The simplification of the calculations needed to construct programs has not resulted in the stripping away of skills. On the contrary, the context within which these processes are carried out is an important factor. All of the process engineers discussed the challenges they face working with advanced machine tools which are capable of manufacturing intricate parts.

Several explained their satisfaction with CAD/CAM work:

> It expands the possibilities of what you can do and reduces the boring part of the job . . . It gives you a lot more time to really apply yourself. It's almost as if, now you're twice as smart because the computer would almost become an extension of your brain.

> I see the methods man of tomorrow being very different. He will have to have a better understanding of the theoretical part of the job. With all of the different types of materials which require new processes, especially in the composites areas . . . a lot of new alloys. They require a new theory [of metalworking].

Methods are responsible for every phase of planning parts manufacturing – from trouble-shooting with manufacturing engineers on the feasibility of production to creating programs, punching machine code tapes, operating tape processing equipment, writing operation sheets and ordering tools for production. The narrow, detailed division of labour in mass production industries is non-existent in this facility. Given the nature of small batch production, the simplification, fragmentation and standardization of the labour process is inefficient. The nature of CAD/CAM as an integrative technology also mitigates against these tendencies. Under different circumstances, CAD/CAM might be used to fragment and routinize job functions. Its capabilities as an information-based tool make large amounts of data instantly available to the user. With the elimination of time-consuming calculations and the addition of software packages that aid in programming, creating a detailed division of labour within the occupation is unproductive.

Traditional arguments on the de-skilling of labour[11] cannot be easily applied in this case. Work has not been sub-divided nor fragmented into semi-skilled tasks. However, given the tendency for automated technology to intensify the labour process, the possibilities for a decrease in the workforce, particularly through attrition, is not unlikely.[12] Given the existing division of labour within the methods department at UFC, the nature of production, and the influence of the trade union, fragmentation and re-organization of the labour process into discrete and repetitive tasks is rather remote.

Although automation has not significantly altered methods' skills and the organization of work *within* the occupation, it has affected the lines of demarcation *between* occupations. One unique feature of automation is the shared body of knowledge which becomes available to a wider range of workers within a firm. Computers provide the user with access to information which heretofore was difficult and time-consuming to locate and practically unusable in its original form. Computerization re-formulates the information and creates an understandable and more accessible framework. A system of commands allows the user access to specific data and functions as well as a context for applying them to the job. The consequences of such an integrated information system has been the elimination of barriers between the different phases of design and manufacture.

Since methods' work primarily involves planning, information relevant to its different phases becomes available to those occupations working closely with them on these processes. CAD/CAM was created to link all phases of the design and production process together. Although UFC has chosen not to automate these connections, nevertheless, information is able to be shared among users in different stages of the process. When asked about the possibility of work shifting between occupations, one process engineer commented:

> I feel that the job we do in our department could be totally eliminated. I'm not saying that the company will run more efficiently without us, but it could be eliminated . . . The work could be split up. Some of it will go up and some will go down. The programming end of it would go down [to machining]. The rest of the work could go up to manufacturing engineering.

With the available data on CAD/CAM, process engineers' role as consultants with manufacturing engineering on the feasibility of parts production could be absorbed by manufacturing engineering. However, process engineers discussed the role of the union in preventing this encroachment:

> There have been cases where they [manufacturing engineering] had tried to do our job. The reason they were stopped is because the union didn't allow it. There were a few instances where grievances were filed [manufacturing engineers belong to another local]. We have an excellent board member who does his job well. I think if it weren't for him, we could lose some of our people.

> The manufacturing department can be bypassed because a design engineer can do the same thing we do. All he has to do, once he finishes [designing] the part is press a couple of buttons and automatically gets the tool pad and, *voilà*, there you have it. The only reason I see that's stopping UFC from going in that direction is because of the union.

Methods know that, realistically, the complexity of the parts produced at UFC make it extremely inefficient – if not practically impossible – to eliminate them completely. As one process engineer notes:

> The drawback to it [eliminating methods] is that it's too inefficient. For one man to really understand tolerances, for example, he has to have an understanding of physics, strength of materials, even chemistry, not to mention basic machining processes, the science of cutting, the types of machine tools and the actual machines themselves. I don't think you could find an engineer who could really do all those real well.

The CAD/CAM system has the capability to access this information. However, the ability of the user, in this case engineering, to apply it to specific – and consistently different – sets of circumstances is questionable. Although all of the relevant data are in the system, unfamiliarity with manufacturing processes limit what design or engineering can actually do with the information. There is a difference of opinion among methods on the extent to which engineering will absorb parts of their work. Grievances filed by the union, however, do indicate that some aspects of their work are capable of being performed by manufacturing engineers.

The more direct and immediate threat, however, is from the shop in the form of MDI (manual data input). The simplification of programming, specifically MDI, gives machinists access to programming without having to learn a programming language.[13] Process engineers also express frustration with simply doing the more functional (rather than theoretical) aspects of CAD/CAM work. Some envision an increasingly limited role for methods to the point where a less educated and experienced person could be trained to respond appropriately to a set of commands. However, as other process engineers are quick to point out, the number of variables that must be considered for each job mitigates against such simplifications.

Its potential for removing a portion of methods' work as programmers is very real. They describe how the process works:

> They've gotten to the point where the machinists can actually take a print and follow the dimension – input the dimensions. So he draws the part in the machine . . . The machine will actually write its own program. At this moment it wouldn't be profitable for the company to eliminate us because most of the machinists don't have enough skill – especially in being able to read the prints.

> Right now – with the equipment I've seen – if you have a man who is smart enough with machining, he can program the machine right there to do just about anything that I can do in the APT language. The technology has caught up to the point that now what they are really going to need is a good methods machinist.

> I guess it's coming to the point where, yes, they are basically taking over our job. As of yet, these machines are still new and they still have a few bugs in them. But the capability is there.

> Also, by the same token, now the machinist has to be a lot smarter. You have to be on the ball and you have to know what's going on. It's almost like a combination of methods and machinist together. In other words,

to get a good man to run that machine he has to be a machinist and a methods man together.

The historical separation of planning and execution envisioned by Taylor becomes increasingly difficult to maintain as work is primarily information-based and readily accessible through the same data base.

As we saw earlier in Chapter 4, planning is essential to machinists' work. Methods are aware of this 'shared domain' in the labour process. Several commented on the problem of ownership of this work:

> I guess the only reason I don't want the machinist to do my job is for job security. I feel they should advance – you can't keep them in the dark ages forever . . . I feel if the company went about it in the right way by saying, 'OK, let these guys do the programming. We have other work for you to do'. But they don't do that.

> They [machinists] do have a little resentment toward us. They feel we hold them back [from programming]. They feel that they should be programming the machines. They feel we hold them back and I agree with them. I don't mind their being trained as long as I'm being trained for something else that's equally challenging.

The Hurco is potentially part of a direct link between design and manufacture in the CAD/CAM system. However, UFC has not integrated it into the system. There is no automatic link up with the design department. Designating it as a conventional machine tool appears to further remove it from becoming a part of a completely computerized production process. Although the bulk of machining continues to be performed on NC using APT programming, the union has treated MDI as a serious infringement on methods' job jurisdiction. According to the union, whether or not methods agree in principle that MDI is appropriately machinists' work, allowing the company to use this capability unrestrained would have set a dangerous precedent. Local 8 brought the case to arbitration.[14] The ruling stated that the company could legally use machinists to manually programme the Hurco. However, they are not allowed to save the program for future use. That would constitute a direct infringement on job jurisdiction rights of process engineers.

The simplification of programming on the Hurco eliminates work from the Methods Department. However, the organization of work and the use of skills for traditional programming jobs remain intact. Methods are not de-skilled, nor do they lose control over the use of their skills. The loss of work is based on the re-definition and removal of work from the bargaining unit.

SUMMARY

The occupation of process engineer is the historical and organizational outcome of Taylorist practices to separate elements of planning and control from the shopfloor and place them in the hands of scientific experts. However, the organizational split between mental and manual work was never fully completed. Although this occupation's primary responsibility is to articulate the most accurate and cost-effective method of part production and assembly, it remains a supportive function to manufacturing. Methods must continue to rely on the cooperation and knowledge of shop workers to respond to unforeseen or unintended consequences of their instructions.

There is no specialization within the methods department at UFC. All process engineers are expected to completely methodize a job and create a taped program when necessary. Given the nature of the company's small and medium batch production and the highly complex nature of these parts, the threat of fragmentation is rather remote.

Within NC programming process engineers have secured undisputed control over planning. The use of a programming language – APT – is, contractually, their sole domain. As a *technical* skill, APT programming carves out for methods a distinct area in planning, fundamentally different from its traditional forms of written, step-by-step instructions. Moreover, the nature of automated machining removes many (although by no means all) of the interventionist practices machinists use to reconstruct methods' directions. Skills used in planning – advanced mathematics, extensive knowledge of machining capabilities, tool usage and metallurgy – are applied in conjunction with the capabilities of the programming language to formulate an NC tape. The acquisition of specific computer language skills provides methods with a distinct role in automated machining. The tapes generated by methods using APT are primarily *machine*, rather than operator, instructions leading to their claim that it is methods, not machinists, who produce NC parts.

Process engineers also use CAD/CAM to generate taped programs for NC machines. Key features of CAD/CAM technology are the translation of abstract principles of machining and complex mathematical calculations into formalized operations, and the use of computer software for problem-solving. It is chiefly an information-processing system which provides a wide range of options for creating a machining program. Methods' use of CAD/CAM is contingent upon their ability to apply this information-based system to their knowledge of manufacturing processes.

The capabilities of the CAD/CAM system have several different effects on methods' skills. First, the system reduces the complex and time-consuming mathematical calculations necessary for creating a part. Second, it provides methods with sub-routines to generate complicated machining operations, simplifying some of the more difficult and complex procedures. Third, the logic of this computer-based information system allows methods to rely more exclusively on a theoretical understanding of machining and expands the opportunities to apply this knowledge to automated machine work.

The application of CAD/CAM does not, however, constitute the development of new skills; it transforms traditional planning skills into a broader application of machining science. The skills methods use are essentially unaffected. Nevertheless, the integrative nature of CAD/CAM poses a potential threat to the occupation's exclusive control over planning information, eliminating barriers between the different phases of design and manufacturing. Given the organization of production at UFC, manu-facturing engineers could absorb some of the less complex programming jobs by accessing the relevant data through the CAD/CAM system. However, the local has diligently fought any infringement on methods' jurisdictional rights to planning work. In addition, the highly complex nature of the parts produced limits the possible elimination of methods from the manufacturing process.

Process engineers do face a challenge to their control over programming through the MDI capability of the Hurco. Although the technology simplifies the method of programming with its user-friendly format, it is management's decision to classify the state-of-the-art automation as a conventional machine tool which establishes jurisdictional rules governing its use. The dispute between union and management over the definition of this machine tool as a conventional or automated machine tool and the corresponding skills required to operate it ultimately was brought to arbitration. It is within this context that Cockburn's third element in the components of skill – its political definition – is most evident.

Management contends that machinists' conventional planning skills are simply transferred to data entry keyboard responses. In addition, specific programming skills (such as APT) which are under the exclusive control of methods are completely irrelevant to Hurco machining. The union counters that all of the tasks associated with programming constitute methods' work. Programmable automation minimizes the importance of specific tasks associated with current job descriptions and, instead, emphasizes the use of informational processes.

Given the changes in the nature of these labour processes, management sought to increase its control over the use of skills by defining these processes in very broad and simplistic terms. The arbitration ruling expanded managerial prerogatives, giving them greater flexibility on the shopfloor and eliminating prevailing work rules regarding specific planning operations.

7 Automation and the Design Process: Drafters

Perhaps the most far-reaching changes in automated work will be in the area of drafting and design. Computer-Aided Design[1] (CAD) was developed in the 1960s with the growing use of computer technology as an information system. The development of the mini- and micro-computers significantly contributed to its application in engineering design and manufacturing.

Several key features of the system make its application to design more complete than it is in manufacturing. Although the design process includes the construction of a visual image, CAD's most important function is to collect and store all of the relevant data on a design. The occupations which share this body of knowledge – engineers, design drafters and drafters – can access parts or all of the information generated in the system.

Moreover, a library of stored shapes and commands provides additional means to work on the data in the system. Unlike machining, which requires extensive knowledge about all of the *physical* motions and operations of a machine tool, design work is primarily an *informational* activity.

With the variety and quantity of data that can be stored and accessed, a user is able to work on a small segment of the design or the entire part. The availability of such a wide range of data to all of the occupations in design work affects the nature, amount of work and use of skills on the CAD system. Traditional barriers between these occupations are threatened as information becomes more easily accessed and applied.

The CAD/CAM system connects the design and manufacturing processes in production, and, according to its critics, minimizes the intervention of manufacturing engineers, process engineers and machinists in the process.[2]

In order to understand and assess the impact of CAD equipment on drafting and design work, one must first understand manual or board work. Drafting is technically known as 'descriptive geometry' – a graphic method of presenting three dimensional problems to determine geometric information. The construction of accurate and finely detailed drawings is a measurement of the extent to which drafters have perfected their skills. They view themselves both as drafters and as apprentices in design work. At UFC, drafters – who are generally young and with limited experience – are trained to detail drawings provided by design drafters and, sometimes, engineers. Detailing involves completing a design with all the specific dimensions that will be needed to produce a part.

Most drafters attend post-secondary technical schools or have a high school technical diploma. Basic drafting education consists of advanced mathematics courses (trigonometry and calculus) and the development of the visual presentation of the design: blueprint drawing. Generally, this means learning how to compose different views of a part in detail and the appropriate placement of the dimensions of the part. Some drafters enter the field through engineering or architecture programmes. Their training focuses less on the conceptualization of different views of a part, emphasizing instead the construction of 3-D models. The difference in approach is significant when drafters begin to work on CAD.

The company hires drafters at the entry level only. Promotions into design drafting are based on seniority and successful completion of a test. In order to learn designing, drafters are sometimes given elementary design work which involves creating a workable design from an engineer's general outline. Design work is, technically, out of classification for drafters. However, design drafters give them the work as part of an informal apprenticeship for promotion into their ranks since the ability to design has been one of the criteria for the job.[3]

The board has historically represented drafting work. Large drafting tables, pristine sheets of paper, drafting or technical pens of different widths and point sizes used for lettering quality, erasers, a straight edge, triangles, compasses and protractors are the tools that comprise the drafting trade. In some sense, the drawings drafters produce are a map for machinists to follow in planning a job. Drafters, as well as design drafters, express a sense of pride and ownership in the quality of the drawings. For most, the blueprint is equivalent to a personal signature:

You can tell my drawings by the lettering – mine's very distinctive. You can pick it out immediately . . . When you're done and you see what you created, it's such a sense of accomplishment. To me, it's almost like a painting.

I like drawing because that's where my talent is . . . To me, the job is line quality and lettering. I feel the sense of craft is how you use your line quality.

You have a lot of different ways that you can present your work. When you look at other people's drawings, you can see the different little things. There's a certain layout of – like, the notes go down to the bottom left, and some people cram everything together, and other people like to stretch it out and make it nice and clear. Some people really get into dark line work. You can always tell the people who are into their jobs because their drawings are perfect. They have the perfect line quality, the perfect lettering. I'm my worst critic. I tear my work apart. It's never good enough. I *love* this work.

For drafters, drawing is both the communication of information and artistic expression. The presentation of work – the layout, lettering and line quality – represent individual style as well as control over the process of producing a blueprint. Blucprints are graphic representations of information for part production. The work is a combination of accurate shape and size information in the context of a visual design. Drafters assess the quality of their work in both areas.

There are three basic stages to board work: (1) conceptualizing the part as a mental construct as well as a series of rough sketches; (2) mathematical calculations for dimensioning and tolerances; and (3) the correct placement of dimensions within the context of a finished drawing. Detailing work performs two functions: in the immediate sense, it provides the shop with necessary information for making parts; in the long term, serves as part of an apprenticeship for drafters to advance into design drafting. Detailing gives the drafter the opportunity to figure out all of the different components of the part and how they relate to the designer's work. One drafter describes how she worked on the board:

If I was handed a basic lay out and told to pull that part out and draw it up, I would first sketch it out myself – front view, side view and top view. And then to get the total picture, sometimes if it's very involved and it's hard to see things, I would do it in three dimensional. That means to do what's called an isometric and I would be more familiar with exactly what's there. After that, I would lay out my drawing, sketch it, place it, lay it all out and see how it fits. Then I'd darken my line work and put in the rest of the dimensions and call-outs.

Because their work is of immediate value on the floor, drafters get a great deal of feedback about the blueprints they construct and the decisions they make in doing layouts. Working closely with design drafters and, often, design engineers, and visiting the shop to see the parts produced give drafters the opportunity to conceptually link the different phases of design and manufacture and to see the relationship of their work to both design and manufacturing.

Within the design department, as within manufacturing at UFC, the division of labour is based solely on occupational distinctions. There are no specialized functions within the occupation where drafters are responsible for discrete parts of the drafting process. Drafting functions are neither fragmented nor routinized. Drafters work on every aspect of the drafting process and exert a considerable degree of control over the detailing operations. In addition, drafters detail highly differentiated and complex parts and are expected to perform all of the necessary operations that go into detailing a job.

COMPUTER-AIDED DRAFTING

The OTA Report defines CAD as:

> an electronic drawing board for design engineers and draftsmen. Instead of drawing a detailed design with pencil and paper, these individuals work at a computer terminal, instructing the computer to combine various lines and curves to produce a drawing of a part and its specifications . . . CAD systems . . . have a library of stored shapes and commands to facilitate the input of design.[4]

As part of an integrated production system, CAD fosters new linkages between different stages of the process. In the case of manufacturing, the diverse aspects of the design and production processes are connected. Blueprint information along with manufacturing data and calculations are used to automatically machine parts. However, within the design process itself, the work is based on essentially the same data. The terminals are all connected to a central computer linking engineering, design and drafting. CAD equipment allows work to be performed on different phases of design using the same images and database. With the use of a lightpen or electronically touch-sensitive drawing board, a designer can produce a model or design and then manipulate the drawing on the screen – adding or editing lines and/or dimensions.

The transition from the board to the CAD terminal alters the physical relationship of the drafter to her or his work. The physical skills associated with board work – sketching out rough ideas and using conventional tools to assist in figuring out the physical dimensions of the part – are eliminated. As one drafter explained:

> You don't get the same sense of your work when you're using a pencil and your mind and hands are working at the same time. Working on CADDS, pushing buttons is a very different feeling than working on the board.

The experience of drafters in the transition from manual to computerized work resembles the machinists' discussion of the comparison of conventional and Hurco machining. The physical manipulation of the tools and the traditional skills associated with craft work are eliminated and the approach to problem-solving which is based on the drafters' understanding of the labour process and is replaced by the logic of an external system. Nevertheless, basic drafting skills continue to be necessary in order to use the CAD system. However, they are now formalized through a set of commands designed to provide the user with feedback on the design. The current organization of the labour process also has a significant impact on the automation of jobs and skills. The degree of specialization, control over skills, and the role of the union in defining the terms of technology implementation shape the definition and use of skills as labour processes are automated.

Most of the drafters are eager to learn the CAD system for several reasons. First, it is state-of-the-art and the future of the occupation. CAD will eventually become the primary source of work for designers and drafters. Second, because the system works more quickly than the board, the more complex designs are given to CAD. The remaining board work tends to be simpler work – mostly change orders – and does not provide drafters with sufficient challenge to develop their drafting and design skills. Third, and related to the previous two points, board drafters who have not learned the CAD system are given less respect and are viewed as less 'professional'. As one drafter put it:

> They (in the CAD room) make us (in the board room) feel like we're working in the stone age – sitting there chiseling things out. When they walk through the room they treat us like we're Neanderthals.

The elimination of hand-practiced drawing skills is a source of concern for some drafters. They are a visible sign of their skill to co-workers and on the shopfloor, as well as the trademark of their occupation. Every drafting textbook emphasizes the importance of letter and line quality and devotes extensive portions of the text to its development.

Just as important (although less visible) is the relationship between conceptualizing a part, determining the dimensions and sectional views, and presenting it in graphic form. Board work is a combination of visual, tactile and conceptual work. For many drafters, proportional representation of a part is obtained visually as well as mathematically.

The automation of drawing marginalizes the work drafters perform on CAD and impedes their chances for promotion into design. Drafters express reservations about the value of CAD for both the application of their skills and its ability to advance their knowledge of the field:

> Basically, they need us to make everything look pretty so the guys on the floor can read them (blueprints). With the terminals, they don't need us anymore. That's why they really don't want draftsmen on it. They want designers and senior draftsmen on it. They don't need you. You're basically worthless on that thing. You don't have to make it look pretty – the machine's gonna do that . . .

> When I came on CADDS, most of the work is done by the designer and I do the polish up work. I don't think I'm learning as much mathematically and how to design as much on the computer as I would on the board. Doing the electronic work, the board is all wired out. All I do is the details for it – the assemblies. Occasionally I do some minor wiring. I've been doing that for the past half year.

> *I* want to be able to create it so know I it's workable. If I get a layout and I have problems with the dimensions – like it's not gonna fit and I can't see why it's not gonna fit – I'm gonna make a little model of it to make sure it's gonna fit. Working out the problem yourself gives *you* the foresight of what it's gonna do. Whereas if you go on the computer, it doesn't give you that. Knowing CADDS is really learning how to use the software to figure out drafting. (emphasis in original)

Drafters face several challenges with the introduction of automated drafting and design. First, there is a real potential for a loss of work. Built into the technology is the ability to create finished drawings with much less time and effort. Designers who have years of drafting experience are capable of completing this process rather easily on CAD.

However, there are factors which mitigate against this process. Designers don't want to do the detailing because, contractually, it is drafters' work. Moreover, this work is the only means drafters have of gaining any experience to advance into design. Since UFC has an internal promotion ladder, there is some managerial investment in seeing that drafters receive some training.

Second, the problem for CAD drafters learning to detail is compounded since CAD operates from an abstract format which transforms drafting information into a drawing by the use of the system's commands. The process of developing a model and/or working through critical calculations aids drafters in understanding what they refer to as the 'how' and 'why' of a design. Without understanding these aspects of design work, drafting on CAD becomes more of a formula approach, using basic commands in the system.

Decisions regarding detailing are increasingly a function of processing information within the system. For design drafters who have already mastered designing, CAD is an additional tool. Drafters, on the other hand, may or may not be able to develop needed design skills working on CAD. Much of their success will depend on the ability to translate basic drafting skills into abstract processes.

Drafters express ambivalence about and have contradictory experiences on CAD. They view it as both a threat to their work as drafters and, subsequently, to their future as designers. However, their mastery of the system (at least those parts relevant to drafting work) and of its possibilities provide a source of encouragement for their ability to access relevant data. This ambivalence is expressed in every interview and pervades their discussions of CAD. What compounds the problem of acquiring design skills is that the more complex parts which could provide opportunities for learning designs are given to CAD. Less challenging work remains on the board.

Since CAD is capable of producing complicated designs with accuracy and speed, time which was once spent calculating and drawing is lessened considerably. Drafters can access all the data that has previously been generated on a particular design. It gives them the opportunity to use the software system as the source of information to complete detailing work. Drafters who have articulated their enormous satisfaction with board work are not averse to CAD work, however. One drafter commented:

> Actually, I can see what I'm drawing a little bit better because I can 3-D it and could make it turn around. Then I could see the whole piece. A lot of times when the engineer hands you the layout, you don't know what it really is. They just tell you to pull it out. On the machine, you can call up his whole creation and spin it around, 3-D it and put it in

color and then you can see the whole outline of your part. That's very good. Then I can see more of what I'm doing and where it's going. Whereas, on the board, that's not afforded to you. You can use all the information that everyone else has put in the machine to help you in your work.

Although all of the drafters felt that CAD's capabilities provide them with additional information and techniques to complete their work, none of them were willing to concede that it does – or should – replace human creativity and a knowledge of drafting. In their interviews, they caution about the results of believing that simply operating the CAD system alone solves some of the complex drafting problems:

I told one of the design engineers that I really wanted to know CADDS and she said that it's not all that it's cracked up to be. 'If you're not careful', she said, 'you can lose some of the knowledge that you should be acquiring along the way'. I guess it's true. Sometimes I feel lazy and don't want to figure it all out. I think, 'Hey, that's what the computer is there for'.

I'm doing basic wiring and printed wiring boards. And I'm finding mistakes that designers make. There are a lot more mistakes done on the computer than on the board. You get lazy with the computer. You start to think that the computer can do everything. But it can't . . . A lot of people just jump right on the computer and start drawing. They just know how to draw on CADDS.

I try not to let the computer do my thinking for me. We have that same knowledge. We just don't think as fast as the computer. People forget that the computer can only work as well as the person operating it.

Design drafters and, to some extent, engineers recognize the limited nature of CAD for drafting work. Information generated on CAD by design and engineering leaves drafters with an increasingly smaller area to use and develop their drafting skills. Some designers have deliberately withheld information or demanded that drafters provide data which CAD could generate. They viewed it as a way for drafters to increase their skills so that CAD would be an aid or tool for them instead of simply a repository of design data to access. Drafters expressed their appreciation of this support. They realize that without such interest on the part of designer and engineers, their skills would not develop.

This guy from CADDS came out to give me a layout to detail. He said, 'Well, here it is. If you have any questions, you know where I sit'. After that, I didn't know if I should go in and ask him a question.

But he said to me, 'I want you to sit here and figure it all out yourself. I want you to be smarter than that computer. I want you to be smarter than the people in there who can just push the buttons'. He really made me work at it to find out all the answers myself. I really liked it.

A lot of the design engineers are really good. They first want *you* to learn it all on your own so that you know it. They said that they have a lot of trouble with the people in the CADDS room now because it's too easy for them. They don't know how to figure things out. They simply let the computer do the figuring out for them.

There's a lot of stuff that the senior design engineers won't put on the layout just so that we can use our brains. So, we have to come up with the tolerance analysis and make sure things fit.

Finishing a drawing by detailing the part trains both the eye and the mind for the processes associated with design work. Drafters have to take into account all of the factors and information provided by design drafters and engineering in order to complete the drawing. Figuring out the tolerances and dimensions and physically creating the drawing of the finished part gives the drafter a grasp of the range of processes involved in and the information needed for designing a part. It is an exercise that provides the drafter with knowledge of and sensitivity to the designing process.

In addition to board work, communication with designers and engineers provide drafters with additional opportunities to perfect their drafting skills and acquire some knowledge of design work. Moreover, their lack of experience in the field and the apprentice-like relationship of drafters to the more advanced design drafters necessitates some kind of informal training process. Design drafters and engineers will often stop by a drafter's desk to check on the progress of an assignment. They point out different approaches to solving a difficult problem or comment on the work. Drafters indicate that the nature of these interactions is significantly changed working on CAD:

There's more team work on the floor [board work] than in the CADDS room. There's more competition inside the room. Some people try to burn others just to get certain work. Even the people who are helping you will try and beat you out. It doesn't happen on the board. I don't know what it is. It seems like, after a while, people start to see themselves more as a computer operator than as a draftsman.

You have more contact with design engineers on the board. In the CADDS room, we're left more on our own. People come by when you're on the board to stop and talk and see what you're working on. They take the time to sit with you and explain how it's gonna work, what it does, what it goes into. In the CADDS room, the design engineers keep more to themselves and stay on the terminals. The terminal itself becomes their world.

The isolation drafters complain about with CAD work is based on several factors. Board work is visually accessible. A designer can stand or sit alongside a drafter and easily view her or his work on the large drafting table. The CAD screen, on the other hand, is much smaller. A drafter would have to get up from the chair and allow the designer to sit in front of the terminal or both of them would have to crowd in front of the screen. The ease of communication which the board facilitates is constrained on CAD.

Moreover, the interactive nature of CAD is often perceived as 'communication'. The system contains all the relevant data on a project which the drafter can access through a system of commands. The user is often viewed as already engaged in some form of learning. Others may hesitate to interfere in what appears to be an ongoing process of communication with the system.

The more complex projects and electronic parts are given to CAD because of its ability to handle the work in a shorter period of time than it could be done manually, if at all, on the board. Traditional board work which historically has identified a skilled drafter is non-existent on CAD. Other, less specific criteria, can be used to assess skill, like the ability to use the available stored data and the command system. Given the relative isolation of the work and the lack of a definition of drafting skill on CAD, competition increases over the available and interesting work in the CAD room.

CAD is primarily an information and graphic system. It provides the user with both faster computing capabilities and a vast data base to store design and manufacturing information. Using a basic command system, users retrieve the stored data and work with it to create a design or complete detail work. The nature of this system fosters new ways of planning a job. Drafters describe their approach in using the CAD system:

I guess with the computer, I store all the information I need to know in that box [computer]. I really don't keep anything up here [points to her head]. If anything happened to that box, that would be it. My friend stores parts in there so that he can just punch them up and put them in. He's creating his own little library at the same time he's working on his drawings. Now, I do it with the books. I save all the pictures and put them in there, with the specs, and I can just refer back to them.

With the machine, you just have to call up the part and place it. You don't have to remember the dimensions or the size or anything like that. You don't store anything in your mind like when you worked on the board. You store it in the machine. That's why I like working on the board, too, because the information stays in my head.

Drafters use the software to build a library of resources they can draw upon to solve problems in detailing. Once drafters have figured out a particular problem, they store the information in the system and retrieve it when faced with a similar project. In this case, drafters are transferring these processes from a mental procedure which they must recall each time they approach the job to a file in a computer system. For experienced drafters, this saves an enormous amount of time in calculations. Less experienced drafters run the risk of creating such libraries without fully understanding all of their implications for the drafting process. Having to recall the procedure re-affirms and re-establishes the connection between a particular body of knowledge, problem solving procedures, and the detailing or design process. One drafter's stated preference for continued experience on the board is an indication of the importance of the need to continue to develop expertise and experience before committing those procedures to the computer's memory.

Working on the terminal itself demands a different way of concept-ualizing a part as well as being able to use the system to detail and design the work. Traditional drafting knowledge is applied through the computer's software command system. Drafters have to translate the mental and manual processes they engage in on the board to a series of abstract symbols whose meaning follows a particular logical pattern. The drafters spoke of the difficulties working with this computer-mediated process:

You begin to lose touch with reality. Sometimes I just have to sit back and think for a while – 'Where did I start from?' Sometimes I just do it by hand first to figure it out, or ask one of the older designers.

Sometimes I'll even go outside to the drawing boards and look around. And sometimes when I go home, I'll think about how to solve the problem.

When you're on the computer it's frustrating because there's something you *know* you can do on the board – no problem at all – but, for some reason, the computer won't do it right. Or you'll hit the wrong number by accident. I've hit a key, instead of making a 1 inch circle, I made a one thousand inch circle. Where did the circle go? Then I gotta look at the screen, zoom the screen down, erase the circle, bring it back up. The computer makes a high pitched sound when you make a mistake and some days you just want to take an axe to it.

Now, with 3-D, you're working with three depths and you can't forget about that when you're moving something. That's what's hard. When you're working on a screen that's in 3-D and you have to think in 3-D. Sometimes, it's the hardest thing to do, especially when you're working with side views.

They have what are called isometric views. As you're inserting a line in one view – say, a front view – you can see it being inserted in the iso. Yet, that can also be very difficult. When you're working on something intricate and look at the iso, you can see every line. It can get confusing. You gotta take your time. Sometimes, if you walk away from the computer and come back, you forget where you were and it takes time to adjust.

The technical capabilities of CAD restructure the relationship of the drafter to her or his work. The ability of the drafter to physically manipulate a drawing, to work the problem out through manually creating a picture and mentally grappling with the parameters of the job is significantly altered with automated drafting. Those processes are subordinated to the drafter's ability to extract relevant data from the system. Because CAD is based solely on information processing, the drafter must develop an appropriate conceptual framework in order to use the system with any degree of efficiency and fluency. The frustration these drafters express points to the conceptual and perceptual difficulties they face in attempting to translate and transform their knowledge into meaningful computerized responses. As Zuboff notes, ' . . . the object of the task seems to have disappeared "behind the screen" and into the information system'.[5]

CAD work demands a different type of conceptualization. Drafting knowledge remains essential – it is not eliminated or simplified. However, drafters need to learn to master the *logic* of the CAD system in order to apply their skills. Several drafters discuss their approach to CAD work:

> Computers and drafting are like two different jobs, even though you're still doing drafting work. When things go wrong with CADDS, it's not really your knowledge of drafting that's gonna help you, but how you can use the computer that's gonna give you the answers.

> I really feel you have to know how to draw *before* you can go on the system. Even if you know how to type in all the commands, you're not gonna know where to place everything. Where your skill lies is if you understand what you're doing with the computer – why the computer is doing it and how it's working. That's where your skill lies on CADDS. But that's just retained in your memory. I don't know exactly how to explain it. You have to learn how to manipulate the system that is there to do what you want.

> After a while of being on the computer, you start thinking like it. A lot of times, that's how I figure out the commands. The computer's very logical, so you ask yourself, 'How would the computer figure this out?' Sometimes things appear simpler than they really are, so you have to be careful or you'll lose something. You start thinking, 'What logical sequence would get me to this point?' After a while, you can feel as if the computer's communicating with you and you start working as one unit. Sometimes it's a little scary – all it is is wires and a screen and some plastic.

The distinction these drafters make between drafting skills and computer logic is an important one. Because they are two very separate – and, indeed, different – processes, the ease with which drafters integrate them is uneven. Learning the capabilities of the computer system is the only means they have to apply drafting skills on CAD. In effect, drafting knowledge is mediated through the CAD system.

Since CAD work is predominantly an abstract process, drafters need to develop and apply a different set of cognitive skills in order to work the system. Their relationship to both drafting knowledge and the tool is transformed. The fundamental work drafters perform remains. Automation rearranges the *approach* to the drafting processes. As a *tool*, CAD alters the techniques used in drafting. As an *information system* it redefines the application of drafting knowledge to the work.

Those who are unable to work within the computer's parameters and according to its logic find CAD particularly frustrating work. Both the method of approaching the problem and the steps in performing the operations are at odds with some of the drafters' own patterns of thinking and creating. In addition, the elimination of manual work removes sensory aids many drafters use in solving problems in detailing or design. However, most of the drafters who work on CAD said that they will often sketch out a part first in order to conceptually and perceptually understand the job. Manually sketching a part provides the opportunity to organize and concretely outline the mental processes associated with drafting and decide on appropriate commands. Many of the drafters said they are reluctant to give up that part of their work. Drawing remains one avenue of control in planning a job and exercising independent judgment.

For drafters, working on CAD is problematic. On the one hand, the access to design, engineering and even manufacturing data provide the opportunity to gain a broader understanding of the entire process. On the other hand, the nature of the interaction with the system and the simplification of complex mathematical processes for tolerances and dimensioning places constraints upon drafters' ability to advance their *specific* kinds of knowledge and skill, particularly as it relates to design experience. Although information about drafting and design is conceived in a more generalized format, the work drafters are responsible for completing is an increasingly narrow part of the process. As one drafter put it:

It's really strange. It's [CAD] the future and you can really learn how to use it to do a lot of different things. But at the same time, it holds you back. It's like you take three steps forward and two back. I feel as if I'm one step behind.

Some drafters experience 3-D CAD work as a further estrangement from drafting. Conceptualizing a part on a flat surface (board) uses particular kinds of mental processes. Drawing the finished part builds upon those processes. Working on a 3-D object created through the CAD system can sometimes cause confusion; drafters complain of losing perspective. One drafter commented:

Sometimes the work gets so clouded on CADDS. It's like a jigsaw puzzle. You're just lookin' for one piece. Sometimes you just have to walk away from it for a while. Then you come back and it's staring you right in the face.

Those drafters who automatically work in 3-D because of their training in art, architecture or engineering, or who began working from 3-D models find CAD much easier to work with. Some drafting schools train students to begin sketching in 2-D and progress to 3-D drawings in creating blueprints. CAD provides an automatic 3-D image which is often a source of frustration for these traditionally trained drafters. For them, working with a 3-D image allows them to visualize *more* details. They use the technical capabilities of the equipment to expand their use of 3-D. Several drafters expressed a preference for 3-D work. They explain how they use the system:

> When I work on the board, the first thing I do is sketch it out on paper. With 3-D, all you have to do is draw a few lines and project them on 3-D. Since it's difficult on CADDS sometimes to visualize the part – even with 3-D – I use different colors to outline different lines on different levels . . . For me, it's easier on the computer. All I do is rotate it around and see where lines are.

> Sometimes on the board, it's really difficult to project a certain view. Where, with the computer, you can play with it and if you give it an idea, it'll branch off from your idea . . . The computer can sometimes take it one step further . . .
> Say I'm looking at the front view of an assembly and it's ten layers deep. You can't see where all this stuff is in line and you want to switch it over into the side view and see all ten layers and you want to make sure you have all the components on the right level . . . You can slice into the computer and say, 'I want to look at level 5'. It'll show you just what's in the area – what you're actually cutting into. Plus you're getting all the three dimensional views.

CAD operates both as a visual aide providing drafters with a detailed image of a part and as an information system which furnishes all of the appropriate data on the job. The extent to which drafters are at ease with this system and able to use it depends upon their ability to approach the problem through the system's logic, format and its visual imagery. Without this understanding, drafters feel they have no control over the system and, hence, over their work.

Drafters who believe that they had mastered the CAD system discuss the increased reliance on mathematical skills in planning their work. Since the machine produces the drawing, those conceptual and mathematical processes associated with *creating* the drawing become a critical aspect of CAD work. Drafters discussed how mathematics is used to construct a drawing on CAD:

I like it because it makes my mind work – because you have to think in your mind – the coordinates, the coordinate locations. Where you're gonna start the line work. It's totally different than the board. It's like a whole other world.

You have to know your coordinates because it's set up on a graph system. When I'm sitting at that terminal and I see the way it's divided into the four coordinates, then all the pre-calc and intermediate algebra starts coming back to me and I realize why I had to know the stuff. I love being on CADDS 'cause I get to use the stuff.

Working on CADDS you do gain more perception because you get used to seeing things in 3-D. With drawing, you don't have to know where the x,y coordinates are. On CADDS, you can't do any work without knowing them. You have to know how the planes work so if you only want to work on a particular side of an object, you have to have the right planes, otherwise it might skip over to the other plane. You have to know how to use them and what commands to use to work with them.

The drafters' relation to the work is transformed by their reliance on more abstract, mathematical processes. The application of drafting knowledge is subordinated to the ability to locate it within the context of mathematical properties. In a sense, the drafter has to develop a new *relationship* to the work. By reconceptualizing the work process around a more mathematical framework, new ways of approaching drafting problems emerge. The system provides additional information and resources to solve drafting problems and requires new patterns of interaction between the user and the data. Zuboff discusses the qualitative change information technology brings to the use of skills: 'Judging a task in the light of experience thus becomes less important than imagining how the task can be reorganized based on new technical capabilities.'[6] Traditional drafting procedures are cumbersome and ineffective unless the drafter is able to place them within the system's framework. As we have seen, many drafters work through this transition by first manually sketching a part in order to 'get their hands on the problem'. Once they feel in control of the *drafting* processes and the information generated from this work, it becomes easier to then submit it to the computer's system.

Drafters who have mastered the more abstract and analytic approach to CAD work use the system to primarily generate *ideas* related to the drafting process, letting the machine complete the image of those ideas. One drafter commented:

You can draw the isometrics after you've constructed the three flat views with one little point and it's there for you. You don't even have to think anymore about what it's gonna take to draw the isometric . . . You could get a lot more involved in what you were doing because you didn't get tangled up in the artistic part. You can just start popping ideas into the machine, faster than what your hands could move on the board. Your brain is really working and you could do that on the screen one, two, three. It left my mind open to be mentally creative. I guess I felt I could be more abstract and scientific about my work. You need a good comprehension of calculus and trig to really work that machine.

Becoming a drafter on CAD means having the understanding and ability to apply the command system to solve drafting problems.

How will this change in the application of knowledge affect drafters in hiring and promotion? Traditionally, drafters on the board have sharpened their skills and gained valuable design experience for promotion into design drafting. Questions emerge in assessing the impact of computer technology on drafting. To what extent will drafters have the opportunity to expand their skills and be promoted into design drafting? To what degree are skills automated and drafting processes routinized, allowing clerks to take over some drafting work as some have feared?[7] On the other hand, do the automated processes permit more experienced designers and engineers to perform the more complex work associated with drafting?

Several factors need to be considered. First, the nature of small and medium batch production and the complex parts produced at UFC make extensive routinization extremely unlikely. The division of labour and the uniformity of products which exists in the mass production firms permit greater intrusion into the labour process, making it easier to further sub-divide and/or automate the work.

Second, the union's ability to thwart the removal of work through filing grievances and acting as 'watchdog' on the floor insure that work is not done out of classification. Although this is certainly a difficult process to monitor, the contract does legally protect the work from being systematically removed from the bargaining unit.

Third, and related to the two previous points, because of the complex nature of the parts and the subsequent difficulty in routinizing the work, it would be extremely difficult for related occupations to assume those duties and eliminate drafting. Nevertheless, there are some changes in the labour process that can impact upon the nature and amount of work drafters will perform.

TRAINING ON CAD: THE ROLE OF THE UNION

The union representing drafters has been an active supporter of automation. The local believes that the union's early involvement in training its members will both increase workers' skills and eliminate the possibility of management being able to shift some of the work to other design occupations, as well as to non-union clerical workers. In addition, because their jobs are a service to engineering – which is expanding at this facility – the union believes these jobs are likely to remain, if not increase. According to the local, access to new technology will only benefit the work they perform for engineering and, therefore, make their members a valuable asset. The president of Local 5 explained the union's position this way:

> Unlike a lot of unions, we're for automation . . . We feel that we have to have our people re-trained, because automation is here to stay. If we keep our people away from it, automation is eventually going to push them out of work. What we're trying to do is – we've asked the company – 'We want the automation, just retrain us. We want to help you and we want to help ourselves. We want to enrich our people'.

The union's decision to pursue training is problematic in its present form. The potential for an increase in the amount of work generated in the area of design does exist. A growing emphasis at the firm on research and development provides favourable conditions for design and drafting employment. Given these conditions, Local 5's support and encouragement for automation is understandable.

Re-training, however, is not an end in itself. Automation is not simply 'another tool' to master, as some in the union have indicated. Nor is it a neutral instrument. For unions, minimizing the technological displacement of workers is an important task. However, ignoring the impact of automation on skills makes the labour process vulnerable to managerial reorganization. Challenging managerial prerogative in this area may ultimately prove to be more difficult. In the context of the present training programme supported by the union, consequences of automation for the *content* of the work is a secondary factor to the issue of maintaining jobs.

The union arranged a deal with the company in which drafters would be trained on the night shift. However, those who are unwilling or unable to transfer for the duration of the training sessions would not receive CAD training. In addition, the training programme is not a separate process from work output. Drafters are expected to produce workable drawings while they learn the CAD system. Most of them complained that this approach was problematic and unworkable. Several drafters commented that the process was geared toward output rather than education:

> I think drafting is destined to disappear with CADDS. It's probably inevitable. They [union] should demand that we get trained well enough on CADDS that we could move up. We need full knowledge of the commands of the machine and of the drafting field. This in-house training is a real *waste*. They should send everyone to school. (emphasis in original interview)

> I basically self-taught myself [on nights]. The training was very lousy. I basically read all the books on commands and used my common sense. There's still a lot of things I don't know, because they didn't train me. They should send you to school.

Being trained how to operate a system and understanding the capabilities of that system are two different and distinct processes. Drafters believe they are being taught only what they need to know to perform their jobs. They feel it will limit their opportunities to advance their understanding of CAD and their ability to become design drafters.[8]

The problem with the programme at UFC is that it focuses on getting drafters to learn the basic uses of the technology. The frustrations these drafters express come not from a resistance to automation, but from their inability to become more involved in understanding the capabilities of the system.[9]

In their support for automation and a training programme, Local 5 argued that increases in skill would inevitably occur. This presupposes, however, that the acquisition of skills is a major component of the programme. The nature of training available to drafters is not specifically designed for developing skills as much as it is for learning to simply operate the system. Moreover, because drafters are relatively inexperienced workers, skills normally developed on the job receive no systematic attention in CAD training. The integration of drafting and CAD work is serendipitous. No real provisions are made for drafters to gain the experience which will give them access to design drafting. One drafter who fears that his chances of promotion into design work are jeopardized with the present training program on CAD said:

> I don't think that the union is *really* on top of the situation. I'm not sure I can pinpoint why, though. I guess I don't see them as being in

control of how automation is being brought in and what it's doing to our work. (emphasis in original interview)

Another commented on the logic of CAD training on the night shift:

I don't know why they're training people on CADDS at night. All the engineers have gone home, so who do you discuss the work with? So, if you have any questions, you're sort of in limbo until you get to talk to the engineer. The people on nights are not really getting any advantage other than learning how to use the computer.

SUMMARY

The implementation of automated technology has contradictory results for drafters. The CAD room does receive more complex parts and demanding work, particularly electronic components which would be nearly impossible to produce on the board. This exposure challenges drafters' knowledge and provides them with opportunities to experiment with different problem-solving approaches. In addition, the integration of information needed to produce a part exposes drafters to a broader range of data and an overall sense of the engineering and design processes. CAD drafters are invited to sit in on reviews where the different occupations responsible for a project – engineering, drafting, manufacturing engineering and quality control – are brought together in a team effort to solve production problems. While it has not, as yet, altered the division of labour, CAD makes the data available on the project accessible to all of its users, making the potential for reorganization possible.

However, the effects on skills are not clear cut. The relationship between automation and the use of drafting skills is quite complex. There are different and conflicting consequences for drafters. On the one hand, drafting skills are increasingly subordinated to the CAD system. Reliance on the computer's capability to perform many of the detailing functions (the core of drafting skills along with manual drawing techniques) makes drafters a less important part of the process. All of the drafters voiced their concern that the more senior and highly skilled design drafters would have no difficulty in completing the detail work on CAD since they already have extensive drafting experience. Most, however, expressed scepticism that the work would become so routinized that drafting clerks, lacking knowledge of detailing, could take over their jobs.

Automation, therefore, does not completely eliminate the need for experienced drafters. The use of CAD, however, may mean a reduction in the number of drafters.[10] US government statistics do indicate there will be an overall decline in the demand for drafters.[11] Even if design drafters do not take over detailing, the speed with which complex calculations can be performed and the elimination of manual drawing can reduce the number of drafters needed for CAD detail work. All of the drafters indicated that with CAD, UFC could eliminate some drafters and continue to produce the same amount of work.

Those who also have developed a more abstract comprehension of the system's capabilities are able to challenge the limits of their traditional drafting skills by using the computer's logic, resources and flexibility. Expansion of these skills is finite, however. CAD does not enable drafters to gain valuable *design* experience. Skills needed for design work are acquired through years of problem-solving and practical application. Since these are separate processes, CAD does not and cannot provide drafters with this information and experience.

CAD drafting has simplified some of the critical skills drafters develop to gain the valuable experiences they need for promotions into design work. CAD offers a broad understanding of design, but under the present training programme, provides little opportunity for systematic and specific training in design work. A possible outcome of this situation could be a two-tier drafting occupation where some drafters can move up into design and others will remain on the board doing some detailing and change orders. One drafter commented that 'there is talk that they'll merge the senior draftsman and draftsman'. If this merger of job classifications is implemented, it will limit the opportunity for mobility of drafters up the ranks to design work.

The routinization of processes of existing drafting work is also an issue facing CAD drafters. The nature of interaction with CAD produces routine responses to the system's commands. The change in the use of skills circumscribes the application of drafting knowledge used in creating detailed drawings and the potential for increases in skills. However, reliance on a more abstract and mathematical approach can also extend the possibilities for solving complex problems, providing drafters are able to comprehend the capabilities of the system.

The implications of CAD drafting are complex and contradictory. What is clear is that the debates on de-skilling and re-skilling need to be reconstituted. The data suggest that an occupation can experience both processes when some work is simplified and routinized, while a level of increasingly abstract skill is demanded. The outcome of these changes in the labour process will depend upon the actions of labour and management to define the nature and use of skills in production.

Automation does eliminate manual drafting skills. The detailing processes which define drafting work and provide the means to advance into design work are seriously curtailed and jeopardized on CAD. However, the vertical integration of skills that is often predicted with computerization is an overly technologically deterministic position. Other social factors affect its impact: the existing division of labour, specialization of skills, the involvement of the union in defining the terms of implementation, and the resistance of workers to perform work that is historically out of classification.

The implementation of automated design technology has different and conflicting consequences for drafters. As inexperienced workers, drafters' skills are gained primarily on the job. Unlike the other occupations which require some previous experience, drafters depend on job training and in-house experience to gain the necessary skills. Manual drafting work provides a wide range of conceptual and problem-solving processes which contribute to both drafting skills and future design work.

CAD work automates and simplifies many of the problem-solving processes associated with drafting. These skills become subordinated to the logic of the computer's information system. The skills subsequently learned on and applied to CAD are circumscribed by the requirements of this system, marginalizing drafters and drafting skills. Moreover, because many of the steps drafters use to solve problems are now part of the computer's information system, some of the processes associated with learning the trade and training for advanced design work are displaced. Although designers are reluctant to perform drafting work, the availability of all of the data on a given part in the system increases the possibility that less skilled workers, like drafting clerks, can access information and perform some of the more routine detailing tasks.

However, the effects on skills are not completely negative. Complex designs are assigned to CAD and expose drafters to a new set of problems and a range of options in producing details of the parts. The relationship between CAD's information system and the use of drafting skills is quite complex. The labour processes associated with CAD work, as described by the drafters, assume a level of prior drafting knowledge and experience. The system is designed so that the drafter applies her or his understanding of drafting processes to the database using a set of commands to access the data and complete the drawing. Drafters are limited in the use of the system by their own knowledge of the field and of the commands in the computer. Those who have grasped a more abstract comprehension of the system's capabilities are able to challenge the limits of their drafting skills by using the computer's logic, resources and flexibility.

Drafting skills and the logic applied to the computer's information system are two distinct processes. As a tool, CAD alters the techniques used in drafting. As an information system, it redefines the application of drafting knowledge to the work. The increased emphasis on abstract mathematical properties in CAD work provides drafters with expanded possibilities in their approach to drafting work. The information and data available through the computer create new patterns for problem-solving.

The union actively supported CAD training, believing that it would secure work in the bargaining unit and provide drafters with additional computer skills. However, consequences for the effects CAD would have on drafting skills and the organization of work were not key issues in the training agreement. The local assumed that learning how to operate the system meant the acquisition of new skills.

The evidence here indicates that CAD makes use of more abstract, mathematical approaches to drafting. Learning commands to access data is not a skill, *per se*, but a highly complex tool for applying existing knowledge and skill to solve drafting problems. This training programme, therefore, is particularly problematic for drafters since they are relatively inexperienced and are in the process of acquiring critical drafting and design skills. These skills cannot be developed through CAD as it is currently taught.

The continued reliance on conventional tasks as a means to delineate boundaries between drafting and design occupations faces new challenges as automation alters the application of skills and information to the labour process. The outcome of these changes will depend upon the actions of labour and management as they struggle over the definition and use of skills in production.

8 Design Drafters

The engineers have been doing schematics on CADDS. That makes it difficult for our union because we have no way of knowing if it's being done and how much of it is being done. It's like we have to have a floorwalker in a five-and-dime looking for shoplifters. These guys are looking for time stealers.

Design Drafter at UFC

Design drafters are the highest ranking occupation within the drafting department. Their job involves working with engineers to create a workable and accurate design of a part or product. Engineers provide the basic ideas and design drafters are expected to design and develop a physical representation of their conceptual work. The *Dictionary of Occupational Titles* defines design drafters as:

> A term used to designate workers who make design drawings to assist in developing experimental ideas evolved by research engineers, using specifications and sketches and employing knowledge of engineering theory and its application to solve mechanical and fabrication problems.[1]

Design drafters are not hired at UFC, they are promoted from within the corporation. Until recently, in order to qualify for promotions, drafters have had to accumulate a minimum of six years' experience and exhibit some knowledge of design work. Currently, requirements have been shifted to the successful completion of a test based on a drafter's ability to solve trigonometry and calculus problems. Less senior drafters who have more recently graduated from technical schools or college have an advantage over drafters with more board experience since their exposure to mathematical formulae and test-taking is considerably more recent.

Although, as many design drafters complained, design work is not simply based on knowing formulae but on the ability to execute a design, the knowledge of advanced mathematics is conducive to the use of the CAD system as we heard in the last chapter. All of the design drafters expressed strong reservations about the wisdom and fairness of maintaining this type of test as a promotional ladder into design work. They believe that reliance

127

on these tests is an inaccurate and misguided policy and are pressuring the union to challenge its validity. Designers discuss this policy:

> In our local, people are against that test. They feel that it is biased in favor of people who are just coming out of school rather than a test of knowledge and experience . . . Some of the younger people that did take the test and have gone up the ladder are complaining because they can't do the job. They feel unsure of themselves.

> On the test, you could get someone who is very good at math, not really know any design, and still pass. They're looking for more technical knowledge. But, when you're a designer, a lot of it is intuitive – taking someone's ideas and actually getting them down on a piece of paper.

Design drafters pride themselves on the extensive knowledge they have accumulated and the control they exert in the labour process over the use of that knowledge. Autonomy and skill in design work is expressed in their ability to create and present a drawing which is accurate and complete and capable of being produced or assembled. Their craft is displayed in the designs they construct and the drawings they produce. Several discuss the role of drawing in design work:

> I like the board. It's such a feeling of accomplishment to build something mentally first, with the aid of the engineer, then build it physically with the pencil on the board. It's my way of communicating what is going on inside my head.

> When I make a drawing, I've been told by the engineers that it looks like I did it in ink. I've been known as 'The Engraver'. When I put a line in, it's there permanently. My drawings make good prints. But it's not only the drawing that's creative. It's coming up with my own ideas about how a part should be designed. Everyone has their own ideas about how to make it – what parts you should put in there to make the designing simpler, the manufacturing cheaper.

Drafting work is judged as a personal signature. The result of practicing one's craft is immediately distinguishable from others' and is viewed as a highly individualized product. For design drafters, this has added meaning. The presentation of work is, in fact, a portrait of technical calculations. Beyond the physical presentation of the work itself are the skills that go into designing a part. Based on conceptual instructions from the engineering department, these designs are layouts of each aspect of a part, its critical dimensions, tolerance studies, and the appropriate format to fit the necessary components together.

The skills design drafters need to be able to generate designs are acquired through years of experience – first as drafters completing detail drawings. As they work on the drawings, drafters learn some of the basic processes that go into creating a workable design. As senior drafters, they are given basic designs and work closely with design drafters, gaining important experience in learning and applying principles of design. Some of the design drafters comment on their apprenticeship at UFC:

> I was able to work with the designers – the class 1 designers – and they gave me parts of their layout to do. If I had problems, I would ask them and learn from them. Unfortunately, for the younger guys, many of these people are ready to retire.

> Knowing formulas is not enough for design work. You have to have a *feeling* of what will work. First you have to come up with a good idea – see if you have to strengthen anything, how things are going to fit and work together. *That* only comes from experience – from years of trial and error.

Just as machinists discussed the importance of 'having a feeling for what works' as the main process through which they approach a job, design drafters also claim it as an important part of their work. The culmination of working at creating accurate and simple constructions of complex designs appears as simply intuitive knowledge. Each of the processes – tolerance studies, dimensioning, layout construction – must be developed, computed and analyzed and ultimately integrated into an overall design. With all of the different variables involved in generating a design, the approach used by a design drafter is dictated as much by his or her knowledge of the technical possibilities as by past experiences.

CAD DESIGN WORK

UFC has had CAD equipment since the early 1970s when it first became available. It was the first company in the area to purchase the system and UFC would often schedule tours for managers in the local area to inspect it. The first system was quite different than the one presently used. It consisted of a basic computer terminal and a plotting table with an arm which would draw the design from information punched into the computer. Two of the design drafters interviewed discuss working on that system:

I started with CADDS 1 where they had the arm that came down and drew for you. It was quite aggravating. They used to have felt tip pens that used to mess up your drawings if you didn't have the paper set down right. It was like an extension of your hand. CADDS 1 was more or less a prototype.

In its early days, CADD wasn't sophisticated enough to know whether there was paper under the pens, or whether the pens had gone dry. Once you told the machine to plot something, it would do it. A guy who started the plot could be expected to plan his next drawing on a side table.

It worked from a tape and the memory would tell it where to go — what points to go to . . . You would digitize it and the arm would follow it and put it where you programmed it. They had breakdowns a lot and if you had a malfunction it would wipe out all the inputs you put in.

The intention of the early CAD system was to increase design drafters' productivity. The lack of a sophisticated printing mechanism and its ensuing problems negated any real gains that might have been made. The contemporary version, CADDS 4X, is a completely different system. The screen provides a 3-D multi-color image and it has an automated print machine.

Among the design drafters interviewed, CAD work is either challenging and rewarding or an unwanted imposition. There are no neutral feelings about automated design work. Those who excel on the system explain their approach to CAD work. They say:

I've spent a solid eight out of twelve years in CADDS. It's just a tool. It's not something that's going to take over your job. The computer can't take over my job because it can't think for me, in the sense of creativity. I don't just push a button like some people seem to believe — especially the bosses — and the work just comes out.

It's not much different from the board to CADDS. I use all the same math. All the thinking processes remain the same. You really have to have the knowledge of design in order to work on it. If you don't have it, you can't work on the machine.

All it is is learning what the machine is capable to doing for you and how to get at that information through the commands. You have to know drafting and design and then know the commands and how to apply them correctly to your knowledge of drafting. It's a challenge to get the machine to do what you want.

These design drafters do not believe their skill or their occupation is threatened by automation. The need for design skills remains a pre-requisite for CAD work. Automation has not replaced the functions of designing. The successful use of CAD depends on the design drafters' ability to translate their knowledge of design into a set of computer instructions.

Design drafters who experience difficulty working on CAD are generally, although not exclusively, senior workers. They explain the frustrations of automated designing:

> If I do a job, I have the front view, the side view and the top view. I can look at everything simultaneously on the pencil drawing. On CADDS you can't do that because of the limitations of the screen. If I'm working on something that is large – a cabinet that is 6 ft. high, it's not gonna come out very good on a 20 inch screen and neither is the details. So you have to get yourself a print. So, you're back to square one. This is the one thing I dislike about CADDS. Maybe if I develop that skill about picturing more, I might be more receptive. I just can't visualize all 3 views simultaneously on that type of setup. It requires a very different way of thinking.

> With the drawings, I could see everything I needed. Now, with the screen, all I see is this green machine which, when I ask it to see a drawing, it's so minute, it looks like one green blob. It can be so frustrating. Sometimes I have to put in commands a half dozen times because I've missed one letter of a word. You have to follow the machine's commands exactly, otherwise nothing gets done.

Design drafters who find CAD confusing have two main problems with the technology: the ability to do all the necessary work in a 3-D image rather than on separate views; and conforming to the CAD format, using the system to input data which, on the board, would be manually calculated, and in that process, conceptualized, and drawn up. CAD work means having to adjust to a completely different set of symbols and responses.

Many of the design drafters who do not experience these difficulties often were trained in other technical fields, such as architecture or engineering.

These schools differ from traditional drafting programs in their approach to creating designs. They train students to work exclusively with 3-D models. This differs from traditional drafting which begins with 2-D drawings before progressing to 3-D work. Building a design through separate views of 2-D drawings develops a very different conceptual approach to design work. This difference appears to play a critical part in the ease with which design drafters adapt to CAD. Perhaps the clearest explanation of the difficulties these design drafters face is expressed by those who have mastered the system:

> A lot of the older designers I see havin' a lot of trouble keep trying to make *drawings* out of the 3-D database and it gets all screwed up. A lot of the designers, especially those who have been on the board for years – they do it the way they would make it on the board. They make a plane view and they call up another view over here and then they try and project from this views over to that view. Then they get confused when it doesn't work. They try to use the same steps they would use to do it manually on the board and apply that to the computer and it just doesn't work that way . . . They try and force the machine to work like a drawing. The single biggest thing I tell guys who are starting to work on CADDS is to think of it as a part instead of a view or a drawing.

> A lot of guys use CADDS to duplicate what they do on the board. They start with the plane and then they look at the side and then they project different things at different depths. When I'm makin' the model on CADDS, I just make the *part*. There is an actual difference in how to approach it. When I design on CADDS, I don't do the actual calling up of the different views. I don't do that until the very end.

> After working with models for a while, I just began to see drawings in 3-D. That's how you spot things that are wrong. You're looking at it as a part and then one view has a line that shouldn't be there. It's really how you conceptualize a drawing . . . Even when I worked on the board, I never really sketched things out. I always began by working in models.

With CAD, the first and probably most important change is in the ability to conceptualize the work in terms of the entire process rather than the construction of separate views. Work is reorganized from a step-by-step process to one that approaches a problem from the perspective of the entire design. The kinds of perceptual skills design drafters have historically developed are altered on CAD. The relationship between a mental image and its physical presentation in the form of views is transformed with the elimination of manual work. The medium now becomes the software – an information system which the designer must access to create a part.

Salzman's analysis of design drafters' skills notes that 'CAD software has the *potential* to analyze the design task to determine (the best choice among a number of procedures'[2] (emphasis added). Potential is a loaded term and has, in the literature, often been interpreted to mean a definitive rather than possible outcome of automation. As Salzman states, the ability of the computer to pre-determine *all* of the decision-making rules would be impossible. It would mean that each and every task would be rationalized into clearly definable steps and the design of a decision-making algorithm 'that computes which steps to invoke, how to execute it, which combinations to use, and the overall consequences of each step and combination of steps'.[3] Certainly, the *theoretical* possibilities of computerized selection exist. But Salzman also points out that skills and decision-making which are developed through experience 'cannot be specified a priori'.[4] The intuition workers refer to or the 'feeling for what works' in a given situation is virtually impossible to subject to the necessary processes of rationalization.

All of the design drafters, regardless of their facility on the terminal, point out that CAD – as an automated technology – will not eliminate the need for their skills nor de-skill the occupation. The basic design processes are still required to generate accurate calculations and images. As several explained:

> It's fairly easy to learn the basics on CADDS – insert line or arc – stuff like that. But to put it together, that's where the experience comes in.

> There's really no difference in the amount of thought or effort you have to put into each job. There's nothing that you can do on the board that can't be done on CADDS. But there's work that you can do on CADDS that you can't do on the board.

> Some of the guys feel that the uniqueness of what you do on the board with your own pencil is lost on CADDS. Actually, it isn't. You can still show that. It's the way you put in a line, where you put the dimensions – things like that – how you separate the views. No two people will do the same drawings on CADDS.

Creating a design through use of the software does not change the essential nature of design skills. The fundamental processes of planning and design work remain intact. In addition, design drafters who have mastered the system claim a certain amount of personal creativity in the presentation of their work. Although the *physical* artistry associated with drawing is eliminated, a sophisticated understanding of the system's capabilities provide design drafters with the possibility of developing a unique style to the drawings generated through CAD.

The design process can be broken down into three main components: (1) obtaining engineering instructions on the nature and specifications of the design; (2) developing a design which reflects these specifications and calculating its tolerances and dimensions; and (3) creating a visual image of the work. Several design drafters explained the process:

> You get an EI [engineering information sheet] which has a list of components and sometimes even drawings of what they look like. Then they give you approximate outside dimensions and what has to hook up in and out of that and sometimes an approximate sketch of how they think it should go. Then you figure out how you're gonna package it, what brackets it's gonna need where, the height – the beginning processes are exactly the same as on the board.

> On the board, I would first lay out the plan and then start to bring up other drawings. You project and show what it's gonna be like in sections. That way you show from above what it's gonna look like and from the side. If you have something that's at a weird angle, you need to know what that's gonna look like in true projection.

> Most of the design process doesn't happen on CADDS. It happens when you ask questions of the engineers. They will tell you what they need – 'This can't be too close to this.' 'We need this much shielding.' 'We need this much room for expansion.' So you figure out in your mind where you want to put the components . . . It's the same process as on the board.

The initial planning phases of the work are strictly design processes. These steps are neither automated nor simplified as a result of CAD. Design drafters use CAD essentially to complete processes they have already developed with their knowledge of design. CAD is not simply a technical process to generate drawings from the data stored in the system. Accessing information is, of course, a necessary component of CAD design. Nevertheless, it is secondary to a thorough understanding of the principles of design. CAD does not replace the need for analytical skills. CADDS 4X is not capable of generating tolerance analyses, for example. Design drafters must continue to perform these functions as they would if they were on the board. As one design drafter explained:

> I still do the tolerance analysis on paper. There are different types of tolerance analysis. The CADDS will just do 'stack up' – the tolerance on this component, this component and that component. It will add up going straight across. But that isn't always the situation. Sometimes you have things that you have to keep in mind in the middle and other reasons where they won't directly stack up and you'll wind up with some other tolerances.

It appears that CAD – at leasts for the present – is limited in the analytical tasks it can process. CAD cannot automate those skills that require judgment and experience, nor those processes that are multi-dimensional.

Although CAD performs most of the mathematical functions and provides a visual image of a part, some of the design drafters continue to sketch parts and calculate dimensions manually as they work on the computer. They discuss the ways in which they use manual operations with CAD:

> I won't think of just putting something down. I figure out something mathematically before I go in, so I know where it is. Then, I use my numbers as a check against the computer's numbers. If they don't agree – I would have made a mistake or a part is in the wrong place. First I measure it out in my model. But if I don't find any mistakes there, I check the base dimensions.

> I still have to sit down and sketch things out. Sometimes I'll work hand in hand with the computer and my hand-drawn sketches. I could sit down and make a quick sketch, go into the computer and create something. I can define it a lot better that way. I can rotate it and examine where I might have some problems. Then, I'll go back to my sketch, modify it a bit and check my numbers and tolerances.

The continued reliance on sketching and calculating is not indicative of design drafters' inability to use the system. Both of these designers have extensive experience on and are proponents of CAD design. Rather, these processes represent traditional design drafters' planning skills which are used in conjunction with the computer.

For design drafters, CAD technology relies on more abstract and conceptual approaches to the work and requires an increased reliance on the theoretical and mathematical properties of designing. Although design drafters also use these skills on the board, automation *systematically* applies these processes in problem-solving and developing designs. This allows a design drafter to explore a wider range of solutions. Constructing a drawing becomes more of an ancillary process to the ability of the design drafter to access the appropriate data from the system. Having a clear grasp of the theoretical underpinnings of design expands the possible available approaches.

There are literally thousands of commands in the system which can be used in many different combinations. However, knowing the appropriate commands is an adjunct to a fundamental understanding of design. The flexibility the system affords also makes it possible for the designer to construct virtually any shape. Depending on the nature of the job, design drafters can produce a design by giving point-to-point instructions, providing the line coordinates, or using a grid to digitize the drawing. The latter allows the designer greater flexibility in choosing where to begin or end the part. Each technique provides the design drafter with different approaches to solving problems. The more sophisticated a designer's understanding of the CAD system, the greater the opportunity she or he will have to use the advanced techniques and commands available to create designs.

The knowledge and experience design drafters apply to a design are similar to the machinists' exercise of judgment, skill and experience in producing a part. Both demand a high degree of autonomy and experience and a wide range of skills to perform their jobs. CAD and CAM are designed as information processing systems. They require that both experienced machinists and design drafters translate the knowledge of their respective fields into a highly abstract form. They must now approach the work from a *theoretical* understanding of the process – a theoretical understanding which is infused with practical experience. This approach reorganizes the ways in which operations are conceptualized and carried out. Traditional step-by-step procedures are nearly impossible to perform as the system assumes a more integrated approach to the labour process.

THE PROBLEM OF APPRENTICESHIP: LEARNING TO DESIGN

Design drafters are concerned that the skill and knowledge which they developed as drafters through extensive board experience is being seriously undermined with CAD. All of the design drafters interviewed expressed grave reservations about the ability of young drafters to acquire those critical design skills working on CAD.

Lately, a lot of the younger people who are in there now (CAD drafters) that haven't had experience in design, sit down and take something I've created and they start drawing and they really don't know how I got there. One of the unfortunate things, too, is that they don't know how to create these things in that 3-D model. They just see the *result* – something that we created – and all they have to do is draw it up. On the board, you have to do more. In that sense, you learn more as a beginning draftsman [on the board].

I think the draftsman is at the short end of the stick. He doesn't know why that dimension is there. As a designer, I lay out a drawing. I put in all my information. Sometimes I don't put all of the dimensions in [on the board]. I feel that the draftsman's gonna scale the dimensions and he has a certain tolerance that he can play around with. He can put the circle a little further out or closer in, a little higher or lower. He's thinking about the design itself. He has to figure out what dimensions to put in. He has to scale it. So he has to *know* what kind of dimension to put in. When I lay it out on CADDS, he doesn't have that latitude anymore. Whatever dimension I put in is what pops up on the screen. I feel it's bad for them. They're not really going to pick anything up from this.

Because CAD is an *information* based system, the data a design drafter inputs to create a drawing results in a nearly complete design. The detailing work that remains will not be sufficient for drafters to expand their drafting knowledge nor begin to understand all of the elements which make up the process of designing. In effect, the detailer works on a smaller portion of the design, learning more about one aspect of the work at the expense of a much broader understanding of the entire process.

There is disagreement within the local regarding the system for training drafters in design work. Designers are sceptical about the collective bargaining agreement reached which established the programme. One of the designers discusses his position on the agreement:

There's been some difference of opinion between myself and the president of the local on this subject. He's been campaigning on getting more people trained on the equipment, which I guess is good in a sense that you can say, 'We have the people to do the work'. My contention always was that we should secure the work and say that it's ours and that it's work that we're hired for and that we *should* be doing. Training on the computer is *another* story.

SHIFTING LINES OF DEMARCATION: THE ISSUE OF 'CREATIVE' WORK

Design drafters are worried about the potential for engineers to absorb segments of their work. The area of greatest concern is in the conceptualization of a part or design, particularly with such work as schematics. Schematics are representations which show the electrical components, resistors, and transistors and how they are wired. They are fairly standard and routine operations which designers consider the 'low end' of their work. Typically, the process begins with a sketch from an engineer. It is then given to designers who finalize all of the necessary details and format it. The distinction between the engineer's sketches and the designer's work is a fine one. Contractually, engineers perform the 'creative' work of design, while the design drafters take the sketch and create a completed image. Technically, engineers' work is a very rough and general idea of a part – essentially, it is a concept of a part they wish to produce.

However, between the two occupations, there is an overlap in the area of conceptual work in this process. It is not limited to CAD work, although the situation is exacerbated with automation. One designer discussed an arbitration ruling dating back many years on this problem:

> There's an open arbitration between Local 5 and the engineers' local going back to 1964. The arbitrator's settlement was left as an *open* arbitration and it still is. The arbitrator said, 'Feel free to re-open it whenever there's a problem'. He set up some guidelines and a design committee which were the reps from the locals and the company to resolve any work jurisdiction problems. As of yet, they haven't re-opened that, although they've threatened to do it.

> Some of the things that were said that were to our advantage were of work of a conceptual nature would be done by our department as well as engineering. What that means is that if an engineer was to take a format or any paper and do a sketch – which is conceptual – he could do that. Once he decided to take that and make a *drawing* of it – that was *our* work.

> Now, with CADDS, he could still do that conceptual work, but now it's done nice and neat. Instead of lettering his text in, he types it in and it comes out the way he wants it.

The original arbitration did not involve automation or CAD. This jurisdictional problem is not strictly an issue over automation. However, the nature of information technology systems such as CAD makes it more difficult to witness work being done out of jurisdiction. Another designer noted:

> Back in the 60s, it was easier to see, easier to differentiate drafting work and engineering work – conceptual and non-conceptual work. Now, it's a much finer line. What used to be called sketching, now can be done on the computer by engineers and it's really not a sketch, but a finished drawing. It is cutting out the drafting occupations, to a certain extent.

CAD allows the engineer to sketch a part or, in this case, a schematic and then automatically refine it into a nearly complete drawing. Since the engineer is essentially doing the same job – that is, creating a drawing, design drafters' claims of jurisdictional rights to the work become an extremely problematic dispute to settle.

With simple and highly repetitive designs, such as schematics or mass production facilities where a more detailed division of labour exists, the labour process is more vulnerable to total integration. However, at UFC, where the division of labour is not as detailed and the range of design work is highly complex, total (or very nearly complete) integration is rather impossible to implement. A design drafter discusses why this situation is not really feasible at UFC:

> The way the system is designed – assuming there's no union – an engineer should be able to put his ideas on the screen and hone them down to a concept that could actually be manufactured, doing all the analysis to make sure it's not gonna fail . . . The engineer would have to learn a lot more. But, you wouldn't want an engineer to produce drawings at his salary. He would still have to learn to draw – a lot of them don't know how. You still have to know the fundamental concepts of drawing.

The role of the union is a critical factor in resisting many of these changes. The 1964 arbitration ruling has, for Local 5, been a precedent for countering such tendencies. However, the lack of critical skills among engineers also directly impinges upon the ability of management to implement such a plan. The theoretical capabilities may exist; but the ability of the engineers to use practical applications of this knowledge is, as we have seen, a very different process.

The problem for design drafters at UFC is the elimination of the lines of demarcation between a creative sketch and a more complete design. Design drafters do not fear that engineers will completely eliminate them from design work. Nevertheless, they do see these shifts as an erosion of work that has been traditionally part of the bargaining unit.

One designer – who is also a union representative – discusses some of the specific ways in which work is transferred, through the simplification and automation of processes, to the engineering department:

> I'd say it's anywhere from ten to twenty per cent of drafting work. Before they would sketch schematics on a pad and have them drawn up by drafting.
>
> There are also engineering clerks, not in our local, who are doing some of our work. The engineering union has multiplied by the hundreds over the last couple of years; whereas our local – which supplies the technical skills – hasn't.
>
> A lot has to do with automation. All the clerks need is a sketch from the engineers. Just by typing into the computer and using the cursor, he could put in boxes, text, lines, schematic symbols – already built into the library. They don't need any drafting background. All you need to know is simple commands . . .
>
> It's something we're constantly fighting. We've filed grievances over it. Unless you have strict contractual wording for work jurisdiction, it's gonna continue to go on . . . Many grievances we filed with engineers' section heads – they were just naive to the contractual agreement. Automation makes it a lot easier to transfer work . . .
>
> To get contract language to protect us is next to impossible these days. I think it would have to be a strike issue over work jurisdiction. I've heard from our attorneys that many arbitrations are lost due to the companies pointing out that the local hasn't suffered from it – people haven't been laid off. Potentially, they could be. It takes a lot of foresight to know what to do about automation and jobs.

The difficulty with developing a concrete policy on automation is the unevenness with which it affects the labour process. While much of the conceptual design work remains intact, CAD automates and simplifies some of the routine functions. This process of deskilling makes them vulnerable to being absorbed by other occupations. Since automated work does not strictly conform to traditional design practices, the nature of work and the context in which it is performed is altered. Devising a strategy for contractual safeguards which address these *specific* changes would have to clearly articulate the vulnerable processes, the way in which work is transformed and specific methods of protecting the work. Given automation's propensity to merge processes, the delineation of affected work is a formidable task.

Designers also express concern over the extent to which other less experienced workers perform different aspects of drafting work, particularly those
associated with detailing:

> We went down to the other end of the building just this week and there
> were two engineering aides – who are in *our* local – doing changes on
> existing schematics on CADDS. They're using EI or sketches to do their
> changes. So, we're getting zapped there, too. They weren't allowed to
> work on formats. They could have brought in a marked up print of
> a schematic or a sketch of a schematic and then *we* put them on a
> format. *We* finalize it and put in on a format.

> Well, the engineering aides can do some of our work. They have no basic
> job training, yet they're doing the schematics. One of the kids we talked
> to didn't even have the CADDS experience in school. He was able to
> learn it there [on the job] and start doing changes on the schematics.

The information stored in CAD permits engineering aides to access enough
data to input the necessary changes. Since the manual processes have been
automated, it is relatively easy for them to complete the work without
having much detailing experience. Nevertheless, their limited knowledge
of the field of drafting and detailing does inhibit them from doing anything
beyond rudimentary change orders. As one design drafter comments:

> An experienced draftsman would know that if you change this – this
> particular view – that it would affect something else. Or, if it's part of
> an assembly drawing, he would know more where to go into the different
> parts and make sure he covers the associated changes. The change orders
> come from the engineers, and maybe the guy don't remember that he
> has to take it off another drawing. Where, the guy in design would
> automatically know this.
>   A clerk might not even realize it. He would be able to copy what you
> give him. You could tell him, 'Take this line out and make this dimension
> bigger'. But he wouldn't have the knowledge to know that it has to fit
> with the next part. I could see somebody up in management, who knows
> nothing about the trade, saying, 'Hey, we could take a clerk and do this
> work now' [on CAD]. They don't know the full ramifications of it.

The impact of CAD on some of the technical aspects of the design
processes must take into account those decisions and processes which

remain beyond the control of the computer system. Neglecting such limitations leads to an emphasis on the inevitability of de-skilling.

In addition to the loss of some work on schematics, CAD also increases productivity on such labour-intensive work as designing printed circuit boards. Design drafters discuss having to cut out paper dollies, manipulate them on the board, paste them on sheets, and create assembly drawings of the printed circuit boards. Jobs which might take four to six weeks on the board are completed in one week on CAD. Design drafters find the work extremely tedious and are relieved that much of it has been automated.

Although design drafters welcome the elimination of the mundane aspects of design, they are increasingly fearful of the consequences of the technological transfer of their work to engineering. According to the designers, the loss of control over work is potentially more damaging to the occupation than the automation of routine work.

Increased accessibility to all of the relevant data and the automation of the physical process of drawing provides engineering with the ability to extend the creative processes associated with design drafting with little additional effort. The distinction between forms of creative work or 'creative' and 'non-creative' design is increasingly unworkable in this situation. One designer, who is also a shop steward, commented on the difficulty facing the union over the issue of automation:

> It's not easy to grieve something like that . . . We would have to come up with some new wording in the contract which would protect our work from being done by engineering. That's the only protection we have. Trying to define creative work is pretty tough. It's all in the interpretation.
>
> The design engineers are not supposed to be doing non-creative work. Where does it stop? How do you know? A guy comes up with something he thinks is creative. He's got to define it a little bit more, so the next guy who gets it understands it. Plus, sometimes, he gets so involved, he just forgets and goes ahead and does the work.

Judgment, experience, access to data and facility with drawing affect the extent to which engineers can and do cross over into design drafters' work. The line between different forms and degrees of design work is increasingly blurred. It calls into question the whole notion of what constitutes creative and non-creative design. However, while the need to clearly define those distinctions for political reasons are great, the tendency of computerized equipment is to merge processes rather than create specialized work.

The question of skill, therefore, is less of an issue for design drafters than is the definition of the nature of the work they and engineers are responsible for on CAD. Those labour processes which are defined as 'creative' are essentially conceptual activities. Attempting to fragment this work into discrete engineering and design elements appears impossible. However, several designers discuss a suggestion by management which might signal a move toward some form of systematization:

> Management wanted us to put down in a book the best way to do the job – whatever it is we're workin' on. Most people aren't really keen on it.

> I think it's a way to systematize or routinize our work. They want to know the 'best way' to do things to increase our productivity. They want the tricks we learned to do in the system.

> I think the union should watch out for the fact that the company wants to set up some kind of system where they find the best way possible to do the job. There's a definite distinction between our work and engineering and detailing. But with CADDS, it's startin' to get a little hazy.

Categorizing the labour process would give management access to both the conceptual work designers engage in and their use of the capabilities of the technology. Using their knowledge of design and development of an understanding of the computer's system, design drafters have, independent of management's direct control, transformed the occupation from a manual to a CAD designing process.

The attempt by management to systematize these decision-making processes may signal a move to influence the use of automation in the area of design. This has several probable consequences.

First, the categorization of design processes can become an indicator to measure productivity. Conceptual work has historically eluded any form of quantification. However, the transformation of abstract processes into a series of computer commands opens up the possibility for at least some of these processes to be documented and, if needed, routinized. Given the nature of batch production, management may not substantially alter the labour process (if at all). Nevertheless, categorization permits a greater understanding of the processes involved in automated design should changes be considered.

Second, the systematization of the processes involved in design may be able to drive a wedge between what is considered to be 'creative' and 'non-creative' work. Processes that are more vulnerable to forms of categorization could become the domain of either engineers or aides or sub-divided among engineering aides or clerks and design drafters. Those processes that demand the expertise of a designer or cannot be sub-divided would remain the core of design work.

Third, systematization can be employed as part of the training of drafters for design work. Drafters, who have little design experience, would have a set of procedures to follow for a range of applications. This process could be used to solve some of the problems of promotion into design and the ability of drafters to handle design work.

The potential for any of these processes occurring, or the extent to which these changes are possible, are mediated by several factors: workers' resistance to participate in this process; the union's role in defining contract language which outlines (to the degree that it is feasible) the demarcation between categories of occupational creativity; and the nature of products and production processes. Decisions regarding work organization are made on all of these levels. The range of options available on the *use* of automation are the results of social negotiations, not technical processes.

In the case of design drafters, the existing division of labour has not been significantly altered by automation. The occupation is also not threatened with the elimination of its skilled work. Instead, the informal delegation of specific tasks to other occupations signals an infringement on the control over some processes which have historically been attached to design drafting.

SUMMARY

Design drafting has been automated for ten years at UFC. Although the earlier stages of CAD were not sophisticated systems, its contemporary application has not proven to be a threat to designers' skills. There have been no substantial changes to the division of labour and the use of skills. The organization of production continues to be based on a broad division of labour, rather than narrow and highly specialized functions within the occupation. The fragmentation and routinization which critics of computerized automation have found, tend to be in mass production industries where occupations are *already* specialized and fragmented. As this study shows, automation cannot be treated as monolithic in the analysis of its effects on the labour process. Wilkinson's case studies of automation in different workplace settings supports these findings:

A variety of organizational forms exists in different firms using very similar techniques, and that the type of work organization which evolves around new technology is dependent on choices made by management which mediate the 'impacts' of the technology. They also indicate that the management strategy developed necessarily takes into account the existing organizational arrangements. . . [5]

Nevertheless, CAD has had an impact on the nature of skills needed to operate the system. Designers are required to develop more abstract problem-solving skills in the construction of a design. In effect, it involves working from a theory of the design process. Designers approach the work through a different medium. The performance of individual tasks is replaced by a broader application of design knowledge to the information technology.

Design skills essentially remain intact and form the core of the labour process. However, both the concrete designing tasks and the method of applying these processes are transformed through computer-mediated symbols. The fundamental changes are in the *approach* to the design process, rather than in the designing skills which are brought to the process.

Design drafters then apply their understanding of design to the CAD system. While it does require conceptual skills, they are not developed in isolation of the design process, but in conjunction with it.

For design drafters, the routinization of work or fragmentation of skills is not an issue or a threat to the occupation. However, while designers are able to further develop conceptual skills, the automation of manual processes and CAD's information storage and retrieval system allow other occupations to perform some of the more routine aspects of designing. This constitutes a form of de-skilling, in that processes which have been subject to automation are simplified to the extent that they can be performed by other workers. In this case, the emphasis resides, not in the traditional notions of de-skilling (that is, direct impact on skills) as in the control over the work. The questions facing designers are: what constitutes design work in general and, more specifically, what are the criteria that delineate the creative processes of engineering from those of designing?

Design drafters are the most skilled and experienced workers in the drafting department. As designers of concrete images of engineering ideas, design drafters exert considerable autonomy over the labour process. Responsible for the feasibility and accuracy of the designs, they apply a range of mathematical and experiential knowledge to the work.

The complex designs of parts, the division of labour and the high level of skill among designers act as constraints against de-skilling. The potential for the fragmentation and narrow specialization of design work which Cooley[6] predicts is unlikely to occur here because of the nature of production and the existing division of labour.

Two of the major issues which confront design drafters are the definition of conceptual work and the apprenticeship of drafters into design which is, ultimately, the future of the occupation. The overlap between engineers and designers around what constitutes conceptual work has historically been a jurisdictional problem. CAD compounds this problem by providing a nearly completed image of the engineer's sketch, eliminating the demarcation between a conceptual and a completed design, as in the case of schematics.

The automation and simplification of some design processes allow sections of the work to be performed out of classification. The vertical integration of routine functions threatens designers' control over that portion of the work and with it, the potential for it to be transferred out of the bargaining unit.

Nevertheless, designers have, independent of management control, transformed the occupation from a manual to a CAD designing process. Because CAD requires a thorough background in design and remains a relatively new technology, design drafters have, over the years, carved out their work as CAD designers. Design drafters have exclusive control over both conceptual design work and its application on CAD.

In an attempt to remove control from designers, management recommended the systematization of CAD design procedures. Such a move could have several consequences for the organization and control of the design process. First, conceptual work has historically eluded quantification and categorization. This process increases the possibility for productivity measures to be introduced and some work to be routinized. Second, it has the potential to divide work more clearly between 'creative' and 'non-creative' tasks, thereby producing conditions for the sub-division of the design process. Third, the reduction of design work into a set of guidelines and discrete tasks simplifies the process, giving drafters access to design work and allowing management to cheapen the labour process.

This accessibility is potentially threatening to the domain of work designers have traditionally and contractually controlled. Designers' skills are not affected, but the technology alters the process through which workers in related occupations obtain information, enabling them to perform some aspects of design work.

Part II
The 'New Industrial Relations'

9 Historical Overview and Review of the Literature

The Prince – 'You're mad, my boy, to go with those people! They're all in the maffia, all troublemakers. A Falconeri should be with us, for the King.'

Tancredi – 'For the King, yes, of course. But which King? . . . Unless we ourselves take a hand now, they'll foist a republic on us. If we want things to stay as they are, things will have to change. D'you understand?'[1]

Labour relations under capitalism have been affected by numerous social forces. The formation of labour unions, strikes, mass demonstrations and political unrest, immigration, and prevailing economic conditions have all had an impact on how corporations have responded to demands for change. The emerging bureaucratic reforms in the workplace – which Quality Circles are a part of – are not new. In order to understand the contemporary debate (and its often misplaced emphasis on industrial democracy and work reform), let us examine the historical programmes and bureaucratic changes which were often linked with the notion of democratic industrial reforms.

SCIENTIFIC MANAGEMENT AND 'WELFARE CAPITALISM'

Two of the most significant movements in labour history which focused on gaining control over production methods and securing workers' cooperation in the process have been scientific management and welfare capitalism – incentive programmes and experiments instituted from the late 1800s through the 1920s.

The latter part of the nineteenth and the early twentieth centuries witnessed a radical shift in the structure of capitalism. The corporation emerged as the dominant industrial form whose growth into a complex organization raised management's concern for new forms of social and technical control. Scientific management and welfare capitalism constituted two major approaches to increase managerial power in the production process and consciously alter prevailing labour-management relations. Managerial control was implemented through the fragmentation of work into simple, repetitive tasks, financial incentives such as profit-sharing and social programmes, and the use of psychological techniques to motivate workers to participate in this new form of work organization.

In the midst of the rapid industrialization of the period, the use of science – in its various forms – to find answers to the 'problems' of control, productivity and labour unrest had enormous appeal to managers. The growing importance of scientific personnel administration and the centralization of production work through the Planning Department made famous by Taylor signified a new era in industrial organization. Scientific management offered capitalists desperate for new methods to increase productivity and control both the *means* to increase production and an ideological *rationale* for fragmenting and rationalizing work.

The idea of a 'rational' (read: neutral) basis for determining production methods appealed to management. The emphasis on a scientific method lent credence and respectability to their attempts at removing workers' decision-making (to the fullest extent possible) from the shop floor to the Planning Department. The tendencies of workers to use their control to restrict output – or as Taylor called it, to 'soldier' – could be labelled as 'irrational' and, therefore, be eliminated through scientific management.

Workers did not view scientific management as 'neutral' or beneficial. The strike at the Watertown Arsenal in 1911, the subsequent House of Representative committee hearings during 1911–12 on the effects of scientific management, and the high employee turnover rates at places such as Ford indicated a less than satisfied or convinced workforce.

In many cases, profit-sharing or other management-instituted social programs coincided with, or followed scientific management (although Taylor believed they were unnecessary if his programme was correctly adopted). 'Cooperation' became the ideological olive branch capital offered labour in exchange for labour peace, union avoidance and increased productivity. As important, it appealed to a general sense of justice and Western democratic ideals. Cooperation would enhance the possibilities that workers (particularly foreign-born) would be more willing to adopt American values, leaving aside national ties, union organizing and the potential for political upheaval.

Profit-sharing plans, most notably the Scanlon Plan, served to further integrate workers into cooperation with corporate goals and profit-making.

With it, management shared a percentage of its increased profits in exchange for stability in production output and a commitment to increase productivity. This served two different, yet related, functions. First, within the industry, production rates could be guaranteed at certain levels, and management could also press for increased rates. Workers would actively participate in setting higher productivity rates. Second, economically and ideologically it provided a mechanism for rewarding workers for their cooperation. Capitalism could be seen as a benefit to those willing to participate in increasing productivity. In gathering evidence on corporate experience with profit-sharing, the Report of the National Industrial Conference Board quoted one manufacturer who said:

> We have every feeling of confidence that in normal times the plan is fully appreciated by the large majority of our employees, but during the present unsettled conditions that obtain throughout the labour world, we are free to confess that nothing short of exorbitant wages has any influence on labor. At the same time, we do believe that our profit sharing plan has had some tendency to hold down the demands, and to counteract the influence of the labor organizer.[2]

INDUSTRIAL DEMOCRACY

During the early decades of the twentieth century the 'thrust for efficiency' and control was not limited to scientific management's shopfloor changes. Labour unrest, a growing immigrant population and fierce corporate competition had created chaos and confusion about the changes brought about by developing monopoly capitalism. Labour unions and 'foreign influences' were cited by politicans as contributing factors to social unrest and capitalists were accused of exacerbating conditions by their relentless drive for profits.

The Progressive Era seemed to be as much concerned with immigrants' assimilation into American society as it was in tempering the ravages of an economic system which appeared to be out of control. The emergence of a movement for 'cooperation' between labour and capital was the reform strategy that Progressives could support and business would control. These strategies appeared to address the worst abuses of capitalism – autocratic control on the floor and growing social inequality. By 'investing' in their employees through profit-sharing and 'welfare capitalism', managers hoped to ensure labour peace. Given the failures of scientific management and the wage incentive system, attention turned toward employee representation plans, psychological motivation and the importance of the Personnel Department as the coordinator and implementor of these programmes.

The heyday for the industrial democracy movement arrived around World War I. President Wilson's War Labor Board mandated the institution of 'works councils' or 'employee representation plans' in critical war production industries, such as steel, rubber and machine production. Pressured to increase production levels, quality and safety standards, and reduce labour turnover and unrest, business and labour agreed to set up these councils. Beyond production output, they also served an ideological role. Progressives would view this as a move toward curbing the excesses of capitalist control by extending democratic rights into the workplace. War propaganda used the fight for political democracy as the rallying cry for American involvement in World War I. Employee representation provided domestic assurance that a return to America's traditional 'democratic ideals' awaited war's end.

Beyond patriotic sentiment and the Progressive's mandate for 'corporate responsibility', capitalism was in the midst of consolidating its control over production. The expansion of global markets and the need for improved methods of coordination generated a movement toward the centralization of administrative decision-making or, what Chandler terms, 'the visible hand of management'.[3] Critical to the growth of companies under developing monopoly capitalism was their ability to coordinate their growing operations, increase productivity and reduce costs. Changes in the management and coordination of production, the development and use of mass production techniques, and the entrance of professional managers – particularly in the financial and personnel departments – were essential to competitive survival in growth industries.

The management of workers and labour relations were an integral part of corporate reorganization. According to Harris,

> scientific personnel administration was part of a long-term trend toward systematization in the use of resources within the 'new factory system'. The objectives of this search for order were higher productivity, lower costs and reduced levels of conflict. Its premise was that the outcome would satisfy workers as well as managers.[4]

Many of the companies that pioneered and heavily invested in social welfare programmes were 'centre firms'. They included: oil refining, electrical manufacturing, public utility, meat-packing and farm equipment industries.[5] This is not to suggest that smaller firms did not invest in these programmes – quite the contrary. However, the move toward *systematic* personnel administration became the priority of the large corporation. More than centralizing personnel functions, it represented an organizational shift toward the centralization of various aspects of decision-making and control. It was as much a response to labour militancy and unrest as it was to the efforts of expansion and consolidation. Not only did it systematize employee relations, but created new forms of accountability, including the rise of the managerial 'expert'. The foreman, stripped of his powers to hire and fire workers, gained responsibility for coordinating production on the shopfloor.

The use of an ideology of 'industrial democracy' needs to be examined in light of the developing changes within industries to control and stifle discontent as management reorganized the structure of the corporation in the period of emerging monopoly capitalism. Ideological sentiments that terms like 'industrial democracy' evoke, deflected attention away from efforts underway to consolidate control within the marketplace and, more directly, over the production process, while simultaneously devising the kinds of programmes, actions and organizations that would facilitate these changes.

The emergence of the CIO and New Deal policies in the 1930s significantly altered labour relations. Union demands for negotiated benefits and job security provisions dominated collective bargaining. In addition, World War II solidified bureaucratic organization, as federal agencies required companies to systematize job classifications and wage schedules.[6]

These developments were part of the restructuring of labour relations which were to define post-War negotiations. Although there were numerous strikes during this period, corporations were willing to negotiate wage and benefit increases as long as markets and profit margins were expanding. Moreover, conflicts between labour and management, which included sit-down strikes and work stoppages, were replaced by legalistic forms of dispute settlements and multi-year contracts, removing much of direct rank and file participation. Regulations surrounding the resolution of disputes and the growing involvement of the government in settlements increased during this period. Management, as well as the unions, had developed extensive bureaucratic structures responsible for handling grievances and negotiating and administering the provisions of the contract.[7]

By the mid 1960s, corporations were facing increased competition from abroad and growing unrest in the workplace. Management began to search for ways to reduce labour costs, eliminate the extensive government regulations and the bureaucratic industrial relations organization that had developed over the past two decades, and encourage workers' cooperation with these goals. Out of these changes emerged various types of QWL programmes, whose purpose was to develop new forms of labour-management relations and establish a foundation for the reorganization of corporate planning, decision-making and control.

The contemporary focus on worker participation is a continuation of interest by the early industrial democracy movement in work group behaviour. Management realizes (as F. W. Taylor had) that the relationships among workers on the shopfloor are a critical part of output restriction. However, as Fraser points out, the human relations programmes currently being implemented require that the work group 'break up established centers of collective resistance and . . . rebuild group solidarities around goals prescribed by management'.[8]

The perspective of much of the research for this reorganization is to identify those factors that promote or inhibit worker cooperation. Concern over these issues was the focus of the 1973 HEW report, *Work in America*.[9] It outlined deep dissatisfaction among American workers with their jobs because of lack of job opportunities and upward mobility, boredom and frustration and reduced productivity levels. Among the suggestions to ease these problems were job re-design to provide greater challenge and the involvement of workers in decisions which would increase responsibility and commitment to their work.

Subsequent research [10] addressed those motivational factors which would engage workers' participation, increase worker productivity and minimize the adversarial relationship between labour and management. Social scientists have focused on attitudinal processes which might increase cooperation. Along with worker participation and the quality of work life, attitudes towards work are often presented within a market research model of job satisfaction. While recognizing the predisposition in human nature toward rewarding and creative forms of work, the experience is essentially transformed into sets of attitudes and needs which can ultimately be satisfied within the context of a 'humanized' market. The social basis for human fulfilment in the work process has thus been shifted toward psychological gratification, leaving aside the context of social relations. In the psychological framework, issues of conflict and control are translated into 'variables' such as technology, management policy, or changing social values, and QWL becomes the answer to *symptoms* of prevailing labour problems – job dissatisfaction, fragmented work, and authoritarian rule.

Nord is critical of this position for ignoring class power in the study of human relations. He states:

Modern human resource management appears limited by the market-oriented system in which it developed and to which it is applied. The impotence of (these) strategies stems, at least in part, from the failure of their adherents to recognize that their own givens are the same forces which produced and continue to sustain the situation they wish to change.[11]

Berg *et al.* also criticize the selective approach QWL's use to engage workers' participation. They state that proponents of such programmes have used the issue of motivation to 'emphasize those aspects of work that *seem* amenable to manipulation in the interests of *both* employees and employers'.[12] This promotes an image of increased worker control over decision-making. However, even supporters of these programmes concede, 'no ultimate control has changed hands'.[13] An in-depth study of two plants represented by the UAW on the extent of control exercised by Quality Circle participants concluded that the notion of workers' control remains largely illusory in that the *numbers* – not the *range* – of choices expanded. As communications devices, they appear to create more openness while, realistically, they ' . . . do not alter the distribution of rewards or authority or the mechanisms through which power is exercised'.[14]

In another context, workers' participation was analyzed in terms of maintaining and increasing productivity. Burawoy[15] studied workers' 'consent' in achieving productivity levels which earn incentive pay. Although he focuses on consent as workers' primary behaviour, he also discusses the organizational arrangements which structure the choices workers have in participating in this process.

Shopfloor behaviour which subvert management's rules and the frictions between workers represent conflict over the organization of production as much as it does consent to participate in the process. Although Burawoy acknowledges the organizational constraints workers operate within and the contradictory processes which underlie their participation, he ultimately rejects these contradictions in order to highlight and focus on the activities of consent.

Resistance and cooperation at the point of production are not simply a matter of individual or work group 'behaviours' but, rather, outcomes of existing social relations which are currently being transformed. Heydebrand[16] refers to the restructuring as technocratic corporatism in which strategies are developed to deal with both long-range planning and crisis management. Aspects of these strategies include decentralization, dehierarchization and increased flexibility. Stein and Kanter[17] have called this reorganization a permanent 'parallel structure' capable of dealing with risks and uncertainty traditional bureaucratic organization is unable to do.

Moreover, Stein and Kanter contend that these structures can also be used to avoid internal organizational conflict. Citing reduced economic growth and limited available jobs for workers, they claim that the parallel organization may be used as part of work reform by: '(1) greatly expanding the job opportunities other than promotion that are available to people; and (2) institutionalizing and developing methods based on temporary jobs and tasks.'[18] Both technocratic corporatism and parallel structure include flexible work organization, rely on problem-solving techniques and independent judgement – precisely those activities associated with QWL programmes and supported by proponents of 'workplace democracy'.

Decentralizing such decisions as scheduling, production costs, and quality control allows – actually requires – information sharing on corporate goals and production development, on the one hand, and production problem-solving and the coordination of activities, on the other. QWL programmes are used to solve such problems because they serve as two-way communication systems. Information about corporate planning is disseminated with an expectation that reactions to these plans will be relayed back to management and that strategies for implementing these plans will be negotiated by workers and foremen. Because this process involves workers' ideas, it is mistakenly viewed as democratic participation.

Kochan *et al.*[19] examine the changing nature of QWL programmes over the past two decades. They present three successive types. Type 1 represents a focus on narrow workplace issues such as enhanced worker-supervisor communication and improved worker attitude. Most unions have commonly referred to this type as 'window dressing' participation. Type 2 addresses broader work organization issues which includes collective bargaining issues. Type 3 expands the previous two and includes, what the authors term, strategic issues, that is, employment security and union representation on management committees. The last model is the most organizationally integrated, describing, perhaps, the 'parallel structure' mentioned above. This typology provides a clearer understanding of the historical shifts in the focus of these programmes and indicates its increasing concern with bureaucratic issues as well as those which have been traditionally under union control as collective bargaining issues.

Wells'[20] study of QWL confirms the limited role workers have in exercising any real authority or power in these groups. He concludes that the less authoritarian *style* of management – not worker autonomy or control – is the *experience* of participation. Once workers attempt to address substantive issues which they believe are important and to propose possible solutions, the limitations of the program becomes apparent.

An issue related to decentralized control is the change in the nature of collective bargaining. Corporations are attempting to abandon traditional bargaining patterns to include concessions in wages and benefits and work rules. Kochan *et al.* point out that although such agreements have occurred historically during other recessions, 'the recent extent of such concessions is unprecedented in the post-Great Depression period'.[21] According to the authors, corporations are looking to expand the bargaining agenda with these issues, pushing for fewer job classifications and greater flexibility. In return, management has agreed to certain job security measures to protect workers against job loss. However, Kochan *et al.* point out:

> the new employment security arrangements protect workers against job loss due to decisions where management has some *choice*, such as outsourcing, the speed at which new technology is introduced, and other efforts to improve productivity. The new employment guarantees do not deal with displacement caused by a general falloff in demand. Thus these programs deal with internal decisions rather than with externally generated processes to reduce employment levels.[22]

These decisions can be plant-specific and part of the process of decentralizing operations. Current managerial efforts to negotiate plant-by-plant rather than industry-wide contracts reflect this tendency.

The issue of cooperation associated with QWL is often extended to the collective bargaining process around concessions such as the elimination of work rules and job classifications. Workers and their unions are asked to participate in increasing the firm's competitiveness. Although these firms may not have, nor plan to implement, QWL programmes, the ideology of non-adversarial relations for the success of a firm (and continued employment) pervades labour relations dialogue. Nevertheless, both QWL and concession bargaining can be understood as part of the process of decentralization, flexibility at the point of production, and as characteristic of emerging labour relations.

10 Quality Circles

Over the last ten years, labour-management relations have undergone numerous changes. On the one hand, much organizational and financial resources have been devoted to various types of QWL programmes. The other trend has been an attack on collective bargaining, eroding past gains made by labour. Demands by corporations for concessions in every area of negotiations – from pension and health insurance to work rules and job categories and assignments – have become commonplace at the bargaining table. Plant closures are threatened as the inevitable consequence of labour's non-compliance.

How can these contradictory strategies be understood? Both are part of corporation reorganization and restructuring. Common aims of concessions bargaining and QWL are the increased flexibility in the use of labour and skills and, ultimately, through these processes, of 'cooperation' and the removal of basic contractual protections, the elimination of labour unions.

Quality Circles are a combination of a neo-human relations approach to gaining workers' cooperation, an organizing tool for addressing productivity problems at the point of production, and part of the process of the decentralization of decision-making. This chapter will address each of these points and examine workers' experiences in the programme and its effects on labour relations.

CONTEMPORARY INDUSTRIAL RELATIONS

Corporate ideology that seeks to explain the contemporary movement for worker participation in such programmes as Quality Circles draws on traditional notions of democracy, updated to reflect current social and economic conditions. Much in the same way that industrial democracy served as ideological support for works councils in the early 1900s, the social and political movements of the 1960s and 1970s – for self-determination and community control among racial, ethnic and women's groups – became the backdrop for 'worker participation' in decision-making through Quality Circles. Current proponents of industrial democracy combine the concern for dignity generated by the movement for civil rights and the need for 'relevance' and self-expression extolled by the 1960s' counter-culture to create a generational interest in various participation schemes. Contemporary strategies in the pursuit of worker cooperation combine ideas promoted by the human relations school of the 1920s – for example, the emphasis on the importance of the informal work group – with the sophisticated techniques of psychological motivation that developed in the 1960s.

Contemporary workers are believed to represent a 'new breed'. Raised in 'post-War affluence', they are said to be no longer satisfied with simply having a job or with the routine nature of most work, as their parents might have been. They are identified as self-seekers who are more interested in intrinsic rewards. But this interest in self-satisfaction need not interfere with increased rates of productivity. In fact, as Yankelovich and Immerwahr suggest, 'expressive values can enhance the work ethic when the people who focus on personal growth see their jobs as an outlet for their own self-expressive development'.[1] They note, however, that 'people are not working as hard as their belief in the work ethic indicates that they should be'.[2]

Faced with a workforce which appears ready, willing and able to increase productivity, American management has been urged by social scientists to stop inhibiting their workers' desire for self-fulfillment and to use this available resource to regain America's competitive edge in the international market.[3] Japan's rise from the devastation of World War II to its current position of economic power is usually pointed out as an example of its efficiency. Labour-management cooperation in general and Quality Circle programmes in particular are often cited as the backbone of Japan's 'economic miracle'. Quality of Work Life advocates also criticize American labour unions, suggesting that 'inflexibility' of work rules and the maintenance of adversarial relations with management also inhibit workers' desire for and opportunity to create meaningful changes in their work.

According to this position, work rules have simply become bureaucratic procedures – part of a previous era of labour-management relations based on management mistrust and their Taylorist vision of the organization of work. With an 'enlightened management' willing to forego adversarial styles of labour relations and cooperate with workers in humanizing the workplace, unions' concern for contract protection appears as a hindrance to what has been called an expanded role for labour in the decision-making process.

Beyond workplace democracy and worker participation as ideological solutions to worker apathy and discontent is a more fundamental reason for promoting Quality Circles. The growing interdependence of a global capitalist economy and the intense competition it engenders has forced corporations to reorganize production methods and internal communications and decision-making, as has been argued earlier. The divisional autonomy along with centralized decision-making and control that characterized industries from the 1920s to the 1960s has become increasingly unworkable as competition forces innovations in products, marketing strategies and production techniques. Centralization of management authority helped control production by increasing volume and standardizing products and production methods; but now shifts in control to a decentralized, yet flexible, interdependent system of management reflect variability of production methods, market changes, and demands for more specialized products and services. Increasingly, however, financial decisions and planning are centralized, leaving the day-to-day production decisions to local or divisional managers.

QUALITY CIRCLES AT UFC

The Quality Circle programme at this UFC plant will be examined as part of the social relations of production. This includes relations at the point of production, at the level of the firm, and, to some extent, within the overall changes in the structure of the capitalist economy.

Quality Circles consist of small groups of about eight to ten workers, a group leader who is a supervisory level manager, and a facilitator from middle level management whose responsibility it is to administer QC activities – training and coordinating Circle activities and acting as a liaison between Circle members and the steering committee. The steering committee is drawn from the ranks of middle and upper management as well as union officials. They represent a cross-section of the major functional areas within the organization, as well as the bargaining units of its members. Some corporations have two steering committees: one, drawn from top management which sets the goals and major policy decisions; the other, mostly middle and first-line management, concerned with specific Circle activities. It appears that UFC has one steering committee which combines both groups of managers, although top level managers will often set the policy and priorities and bring them before the steering committee for implementation. Workers are recruited from the same department or across departments where they are involved in different stages of producing the same product.

UFC has had QCs at this facility since 1981, with anywhere from 12 to 22 operating at one time.[4] The variance in numbers reflects both the disbanding of Circles and the formation of new ones so that each Circle is at a very different stage in the process. They involve blue and white collar, union and non-union workers. They exist in areas which are labour-intensive as well as those which are automated or in the process of automation.

Proponents of QC claim two very different goals of the programme: (1) increased quality and productivity; and (2) increased worker satisfaction. In the first *Quality Circle Report* published by UFC, management states:

> The Circle members are applying their experience and creativity to analyze and solve problems affecting their work. They're putting more of themselves into their jobs and getting more satisfaction out of their work. All of us spend a large part of our adult lives working at our jobs. That time should be well spent, it should be meaningful and it should be productive.[5]

This general, yet laudatory, endorsement of QC makes a very direct connection between productivity and job satisfaction.[6] The seemingly innocuous statement is, in fact, at the heart of the debate over the meaning and role of QC for workers and management, and its relationship to reform within the workplace.

Those who applaud its emergence point to its democratic structure of decision-making, the sweeping away of the traditional autocratic rule of management which has historically plagued labour relations, replacing it with 'enlightened managers' who recognize the contribution workers make to the production process, and the 'self-actualization' of workers' potential through their participation.[7] Those who reject such schemes as QC view them as cooptation of legitimate worker discontent, superficial treatment of more deep-seated organizational problems, and as a means to instill a more compliant attitude among workers toward increasing productivity and efficiency with little actual participation in decision-making.[8] Other critics add that such programmes are primarily a response to increased international competition which is forcing corporations to seek new ways of increasing cost-effectiveness.[9]

Infusing and, in some sense, structuring the discussion of the role and importance of QC is the issue of industrial democracy. By focusing on the notion of worker participation in some form of decision-making process or work reorganization plan, the discourse is contained within a framework which draws on the very notion of democratic tradition. The use of key words or phrases which symbolically embody this tradition – participation, self-expression, choice and control – are applied to the nature and substance of such programmes.

The appeal is a type of economic nationalism where the stated values of US society (that is, freedom, democracy, and individuality) are presented in the same context as the need to reorganize capital in order to improve the corporation's (as well as the nation's) competitive position in the world market.

As corporations restructure their organizations by streamlining operations, expanding into multi-product companies and automating production, the coordination of these activities becomes a central issue. Predictability takes on added importance as long-range corporate and financial planning are centralized and day-to-day decisions and 'crisis' management are decentralized. Creating an organizational mechanism based on specific forms of cooperative relations provides management with an additional method of exercising control over the market and over production as it reorganizes its bureaucratic structure.

UFC has moved toward such a reorganization. It has consolidated its five distinct companies around two business segments – electronics and machinery products. UFC subsequently sold off one of its machinery producing divisions, focusing more directly on the electronics industry. The corporation has also centralized its business and financial planning. The Chairman of UFC describes these changes:

> In the past, we took the separate business plans submitted by each of the five operations, massaged them a little, and then bound them together to call it our corporate plan . . . Now we're providing leadership. We're getting more involved in the planning process and we're taking a long comprehensive look at our combined resources and market directions . . . [10]

He further states that additional committees were formed to coordinate these changes: the Electronics Executive Committee and the Machinery Products Executive Committee. In a subsequent interview, the Chairman discusses how these committees will operate:

> They will be the focal points for our business and marketing strategy, long-range planning, product development and productivity improvement . . . We have to make the organization more integrated, more efficient and more effective. The committees are designed to achieve

those ends . . . without interfering with the ongoing co-operation of the line organizations.[11]

The article notes that additional changes will be made, including layoffs, salary freezes, plant shutdowns and consolidations. All of these measures contribute to the corporation's restructuring plan. In response to the reporter's question that it seemed like a description of centralized management, he stated:

> I'd call it 'greater corporate strategic control'. But if you want to call it greater centralization of our direction and increased consolidation of our resources, that would be accurate . . .
> The executive committees will establish the overall strategy for UFC. They also will oversee the planning, approve the major actions and control the basic course to be taken by the five operating units. But they won't get involved in the day-to-day activities of the operation.[12]

According to the article, these committees will report to a new Vice President for corporate planning and business development whose strengths are in market trends and business direction rather than product areas. In a subsequent article on a meeting of the divisional managers at corporate headquarters, the Chairman was quoted outlining what these changes will mean for the different divisions:

> We have restructured our planning in terms of people, organization and procedures. In the past, our strategic plan was, generally, the sum of the business units' strategic plans. Corporate management didn't get involved until very late in the game. Today, the corporate planning people and other corporate executives get involved at the outset. This helps us identify issues – together with the business unit – from an overall corporate perspective. And it also accelerates the decision-making process.[13]

The creation of executive committees for each major product area and their accountability to a new corporate planning and development department seems to represent three major shifts within the corporation: (1) efforts at greater coordination among interdependent operations at the executive level; (2) the consolidation and centralization of financial and overall corporate planning ; and (3) the decentralization of decisions concerning day-to-day operations. It also appears that local managers will be responsible for implementing the goals which the committees set while, at the same time, have increased autonomy in the actual operations of their facilities.

This move by UFC to restructure corporate planning and decision-making processes supports a trend within corporations which began in the 1960s. New mechanisms for communication, cooperation and control such as the ones mentioned here are designed to minimize the risks associated with the changes taking place within the structure of capitalism. As Stein and Kanter point out, the world economy is likely to demand that those

> bureaucracies respond to turbulent environments of high uncertainty, rapid change and permeable boundaries. The oil crisis, growing inflationary pressures, market uncertainties due to foreign competition, regulatory constraints . . . organizations . . . face the need to respond more flexibly and rapidly to these environments by solving a continuing series of new problems and changing their traditional internal focus to a more external one.[14]

The nature of these global economic changes have demanded a new form of bureaucratic response. The kinds of reorganization taking place at UFC provide a vehicle for reducing uncertainty and asserting new mechanisms of control in the marketplace. However, organizations are not responding only to external economic changes. Labour demands and conflict also contribute to the attempts to minimize corporate insecurity. Stein and Kanter argue that these internal pressures are linked to a slower growth economy.[15] Its outcome and impact on the organization will include:

> a demand for more opportunity (career progress and the chance to develop) from a wider range of employees, including growing numbers of educated workers . . . for more power (a sense of entitlement, more rights, and job autonomy and less 'over-supervision').[16]

What Stein and Kanter suggest is that it is a stagnant economy and the resulting limitations on workers' opportunity and mobility – and not, primarily, the need for workplace democracy – which are responsible for demands of greater job satisfaction and resistance to traditional bureaucratic inducements for increased productivity. This is a critical *structural* explanation for the implementation of QWL-type programmes. It stands in sharp contrast to those who explain the 'phenomenon' (as if it were a static entity) of workers' discontent in terms of both generational and traditional values.[17]

Historically, workers have always resisted routinized work and the loss of control over the labour process. To suggest (as most proponents of QWL have) that demands for job satisfaction and participation are responses of the 'rising expectations of the new breed of American workers' is to discount historical movements around industrial democracy and worker control. Moreover, these analyses tend to begin with observable symptoms of the outcome of conflict rather than structural explanations of demands for change.

IMPLEMENTING A QC PROGRAMME

The most popular format for providing the emerging corporate structure with increased flexibility is the QC programme (or its variations, Employee Involvement, Job Enrichment, Labour-Management Cooperation Programmes). QCs function as an additional mechanism for solving production problems, introducing corporate goals and strategies on the local level, as well as to provide a means for some form of participatory decision-making. As a 'parallel structure' within the organization, a QC programme has three major objectives: (1) that it develops goals that are compatible with the overall goals the corporation has set for itself; (2) that it addresses some of the crisis issues facing corporations within an increasingly competitive world market – quality, productivity, coordination and product changes; and (3) that it provides a process within which workers can participate in selected areas of bureaucratic decision-making.

Although traditional bureaucratic rules, procedures and decision-making processes are suspended in QC operations, it is not an indication of a lack of a formal organization. QC has its own formal structure. As Stein and Kanter indicate, previous bureaucratic organization is not replaced. 'The parallel organization provides an *additional* management structure to that structure which already exists'[18] [emphasis added]. The Members Manual used at UFC stresses this point. It states:

> Caution must be observed to avoid creating another entity within the organization. This is not a movement designed to introduce radical organizational change, rather it is a powerful tool to help make the current organization work better.[19]

The extent to which this structure becomes rooted within (or operates alongside) the existing bureaucratic framework is contingent upon several factors: (1) its support by *all* levels of management – from corporate executives to the foremen; (2) the cooperation and participation of workers and the union (if one is present) in the process; and (3) the outcomes and effects the decisions reached within these groups have on the labour process and the workers.

Let us now turn to an examination of the structure of the QC programme at UFC's facility at Pine Hill.

Decisions regarding the establishment of QCs are made in a steering committee composed of managers and union officials. One of the design drafters, a former union official, served on the committee. He describes the steering committee at UFC:

> I joined the steering committee because I became vice president of the local . . . Starting Quality Circles at the company level was kinda clumsy . . . The people who were put in charge, I feel, really didn't want to be there. They didn't want to do it. These were managers – one of them was the vice president's right-hand man.

> I think they were forced into it by the corporation – it came from headquarters. Nobody *really* said they were against it, but that was the feeling I got.

> The chairman (of the steering committee) was a typical, arrogant type of middle management. He ran the thing like a typical manager. *He* decided how many facilitators to have, how many circles to have, what public relations type things they should do.

The role of the steering committee is to set the policy and goals of the QC programme on the local level. Upper management at the facility remain firmly in control of instituting the programme and of making key decisions on its structure and operations. The involvement of lower managerial personnel remains, to a great extent, a traditional one – to carry out policies and procedures.

The locals agreed to participate, though with a great deal of scepticism. Those leaders who seemed the most enthusiastic saw it as an opportunity to gain job security for their members through improved quality and productivity increases. To some extent, they felt that members might also gain self-esteem from directly participating in decisions concerning their work. However, even among enthusiasts, there was a sense of caution about supporting QCs.

Clear guidelines were established which made collective bargaining issues strictly off limits. Union representatives were to attend the meetings to insure that no union issues were discussed. As we shall see, this mandate was extremely difficult to uphold. Cole[20] states that such mandates are 'convenient fictions' and that it is nearly impossible to stay out of contract issues where productivity is concerned.

Quality Circles were formed in a number of different areas, among them engineering, financial, assembly, design and methods. Although UFC promotes the expansion of these programmes throughout the corporation and at this particular facility, not all departments are designated to have them. Machinists, for example, were told by their supervisors that there would be no Quality Circle in the department. One of the shop stewards describes the response machinists received upon inquiring about the possibility of a Circle in the shop:

> We were refused a Quality Circle – turned down flat. UFC had a campaign goin'on about them. The Navy came around with a movie about 'em. It got some of the younger guys in the shop all psyched up. They went to management and said, 'We want a Quality Circle'. Management said, 'Forget it. You're not going to have one, that's it'.

From the point of view of the more senior machinists, they felt there was little Quality Circles had to offer the shop.

> I think the shop is reasonably efficient. There's not too much they can ask us to do. After all, we have to figure out the best way to do the job, anyway.

> As far as how something can be done better and cheaper . . . and I think that is the purpose of Quality Circles . . . I think this comes into play more in areas where you're working with a process that involves a chain of steps. Things in the shop are more straightforward than that.

Training

Workers in areas designated for Quality Circles disputed the claim that participation is completely voluntary. Some members said they were chosen by supervisors to participate even though they had not expressed any interest in the programme. Others who had been interested were actively discouraged or simply not chosen.

Training sessions ranged from a complete course – weekly for three months – to a one week session, or no formal training at all. It appears that workers who signed up within the first year and a half of the programme received the most thorough training.

The training programme is designed to have two effects: (1) to instruct workers in the particular types of problem-solving techniques which will be used in the Circle; and (2) to focus on the primary goals and objectives of the programme – to identify those processes which can improve productivity and efficiency, and to reduce costs. As the manual states in its introduction and refers to throughout each stage of the training: ' . . . talk in those terms that are important to management – reduced defect and scrap rates, lowered costs and schedule improvements. A good presentation will also emphasize harmony, teamwork and cooperation.'[21] These goals are designed to both identify problems that are of concern to management and encourage a less adversarial approach in finding solutions.

The goals of the training sessions are to familiarize participants with group dynamics and decision-making processes, and to prepare members for problem identification, analysis and solutions using specific measurement techniques. The participants' training manual provides excellent data on how these processes are carried out and on the goals of the Quality Circle programme.

A major emphasis in the manual is learning how to 'problem solve'. It stresses the importance of group solicitation of ideas through 'brainstorming' sessions and methods of data gathering and analysis. The manual provides numerous examples of 'appropriate' problems and factors which contribute or 'cause' them. Every one of them focus on quality, cost savings and productivity.

Workers are taught sampling techniques and measurements and are instructed in methods of gathering information on the area(s) identified as problematic. These processes are explained as part of group decision making processes. The programme stresses the importance of *group* responsibility for generating ideas and problem solving. Training includes some role playing and exercises that emphasize communication skills and group decision-making. One of the more interesting exercises mentioned by participants can be termed as 'crisis decision-making'. A former member describes the exercise:

They gave us one situation to figure out. They said, 'What if you were on a boat'. Then they gave us the details. There were three people altogether on the boat. They gave us a list of things that were on board.

You had to get rid of certain things. In what order would you get rid of them and why.

We had about six people in the group and we had a half hour to put them in order. You had to convince the other people why you chose those things. And everyone had to agree on them. On the list was: a mirror, chocolate candy, ham, a half gallon of water, a gallon of gas and oil mixed, a book of matches, a sextant, a transistor radio and marine maps. You're a 1,000 miles in the middle of the Pacific Ocean with no land in sight.

Well, you start rationalizing – you don't need the maps 'cause you don't know where you are anyway. The sextant is only good with the sun and if no one knows how to use it, you might as well throw it away. Believe it or not, what do you think the most important items were? Water, the oil and gas, and matches. You could ignite the stuff and throw it overboard.

This process fosters several behaviour patterns. First, it structures the group around short-term crisis situations. Topics are chosen based on immediate concerns. Second, by emphasizing 'lifeboat' conditions, choices have to be made regarding the elimination of certain items or processes. Survival depends upon the ability to order priorities. Third, consensus decision-making binds participants together both in the process of choosing appropriate actions as well as in the conclusions they reach. Parker, in his book, *Inside the Circle*,[22] identifies yet another strategy. He points out that this particular problem solving exercise encourages workers to see QWL as a mutual benefit for labour and management – 'a "we're all in this together" philosophy'.

Quality Circle participants are trained to identify organizational issues which might bear – to some extent, at least – a resemblance to this lifeboat scenario. Guiding these decisions, however, are traditional bureaucratic concerns: productivity, efficiency and cooperation. The exercises for data-gathering and problem solving-emphasize production problems, namely the elimination of defects. Addressing the issue of quality appears, on the surface, to be an issue shared by labour and management. Workers have their own standards regarding quality output and have traditionally (where it is possible) used their own initiative and methods to produce good parts. By focusing on this area in the initial stages of the programme, confrontational issues are avoided.

However, a closer analysis of the data-gathering process reveals another side to Quality Circles. A key element is deciding how to gather and organize the data. Once a major problem has been identified, the manual presents a model for locating possible causes. Four major processes are outlined: manpower, methods, material and machines – what is known as the Fishbone Diagram. Issues concerning the use of labour power and production methods are clearly union issues. Given this format, it is extremely unlikely that contract issues can be avoided, even with the best of intentions. The potential for this problem to emerge in Quality Circle sessions becomes more obvious as additional information needed to solve these production problems is needed. For example, there are several options listed for finding the causes of defects. They range from identification by machine, type of defect, shift and employee. This is a crucial step in the Quality Circle process because it is a point where workers begin to address issues around the organization of work as well as work resistance and control exercised at the point of production. It appears to be the area with the greatest potential for conflict or cooptation within the quality circle process.

One of the participants discusses her experiences in the training programme:

> To me, the whole training process was an attempt at brainwashing. We would discuss something and they said, 'No, you're looking at it the wrong way'. Or, they would say, 'No, that has nothing to do with Quality Circles. We're going to talk about this . . . ' They would channel whatever we had to say to fit in with what they wanted us to do. It was all from their point of view.

Training members to focus on particular topics and processes is a format for learning Quality Circle techniques in problem solving, and as a means to limit potential conflict around the issue of 'appropriate' Quality Circle ideas and problems.

The notion of ownership of ideas, processes and solutions is also part of quality circle training and a potential source of conflict. Successfully integrating participants into quality circles means also providing them with some means of identification and control. As one member states:

> The thing they tried to drum into us the whole time is that this is *your* circle – you do with it what you want to do.

The opportunity for collective responsibility for change on the shopfloor is a powerful incentive to participate. Ownership and participation are not neutral terms, but have political meanings attached to them. How workers interpret their meanings can influence the kinds of issues raised, the solutions that are proposed and management's response to them, and the 'life span' of the circle. The following sections will address these issues as the operations of the circles are described and analyzed.

QUALITY CIRCLE PARTICIPATION

Perhaps the best place to begin to examine quality circles is the most obvious – why people join. What issues were important to them? What did they hope to accomplish? Several members discuss their decision to join:

> The union vice president at the time thought it was a good idea for me to join, being a rep in the union. There were things I wanted to get done and I thought it would be a good place to do it.

> It's a good experience to get involved where you work. It's a forum to voice my opinion on certain things. It gives you a way to control the situation.

> The idea sounded good. They told us it was a means for us to eliminate some of the problems that held us back in the job. *I* saw it as a way of speaking my mind. (emphasis in interview)

> UFC wanted to create them and gave us a lot of literature telling us how wonderful Quality Circles were. Because of the exposure I've had with the union, I felt that I should give it a try. I saw an opportunity to possibly learn something.

While some of the members have specific ideas and issues to explore in quality circles, and others express vague notions of involvement, they all believe that participation would ultimately benefit themselves and their work groups.

Nearly three-quarters of the participants are active union members, union representatives or former union representatives. Those who were not active members still defended the role of the union as their sole representative to the company. Quality circles were not viewed as a replacement or substitute for union activity or representation.

Participants view quality circles as a means to contribute to making their work area safer, less disorganized and, in some sense, to exert control over their environment. Participants believed that joining a circle could eliminate some of the chronic complaints they had about their jobs which might not necessarily be covered by collective bargaining. One participant explained it this way:

> I still want the union to protect my ass, to go toe-to-toe with management and fight for what's ours. But here's an area where I think I can have an affect on what happens on a day-to-day basis.

Those who did not join were sceptical, if not outright hostile to the programme.

> I'm not involved in the quality circle. I don't believe in them. I think they're bullshit. Giving them all the information – OUR information that they ask us for . . . 'I'm not going to do his (supervisor's) job for him. That's what they want. They're not pushin' for anything major. It's penny-ante bullshit. (emphasis in interview)

> I think the people who join initially have good intentions. It's the company's way of making someone feel good about having to come here to work. '*I* advised a major corporation'. They're taking his ideas and he's not really getting anything in return except to 'feel' good. It's cosmetic as far as any kind of decision-making is concerned. It's just another ploy. They're there for another purpose. They're not there for workers' needs. (emphasis in interview)

> I guess I kind of see it as a company candy bar on a string. They ask for your ideas and only give you what's cost-effective to them. I think they're could be benefits in them, but it's a loaded benefit . . . I feel it's a company tool. They're gonna get something out of it – we aren't.

> People joined because it was something new. As a union rep, I just took a 'wait and see' attitude. Sometimes it's better just to let it die on its own. The workers see that it is nothin' but a big scam. It's its own propaganda.

A basic mistrust of management's intentions led these workers to reject participation in the programme. However, their scepticism does not support proponents' views that QCs are partly a response to generational attitudes toward participation. A more plausible explanation for these senior workers is that exposure to management's past practices – not a lack of interest in their work or workplace democracy – gave them little reason to believe that there would be any significant changes in management's behaviour in the Quality Circle programme. As Goldman and Van Houten point out, the notion of the 'changing worker' is an historical outcome rather than a generational or cultural phenomenon.

Even aside from the issue of worker alienation, it is likely workers' motivation today differs from the past. Unions have won enough strength and dignity for unionized workers so that they are less amenable to fear as a motivator. At the same time, workers are more aware that hard work will not necessarily lead to advancement.[23]

QUALITY CIRCLE ISSUES

The first project every circle tackles is the choice of a name for the group. The name is supposed to reflect the occupations of the members and the work they are engaged in, providing a 'group identity'. Methods' circle was called 'Checkmates', design drafters were 'New Dimensions' and 'The Unparallels', Assembly called themselves 'The Challengers', and Sheet Metal, 'The G-Men'. This naming process serves two purposes. First, it provides a relatively quick, 'successful' group decision, proving that the quality circle process does work, and an issue that is essentially 'conflict-free'. Second, it provides members with a separate identity, apart from other workers on the floor and from union affiliation. It encourages group solidarity both in the successful completion of a project, and in the shared experience of creating an identity. Aside from creating a particular group dynamic and identity, deciding on a name as the initial 'task' of the circle clouds the difference between the meanings of participation within and of control over the decision-making process. In Witte's study of QWL programmes, he draws a clear distinction between these two processes. According to Witte, control refers to 'the ability of an individual or group to determine unilaterally the outcomes of a decision making process.' On the other hand, participation is defined as 'actions by an individual or group that affect outcomes in a decision-making process'.[24] In actuality, control over decision-making and the power to institute change in organizations remains in the hands of management. The decisions made by the Quality Circle steering committee and the conflict within the circles over 'appropriate' topics and proposal approvals are an indication of the struggle to exert control within quality circles by participants.

A study conducted by Robert J. Thomas[25] on quality circles in the auto industry, found that facilitators encouraged workers to address particular problems, especially those around improving working conditions. The rationale behind this approach is that changes like more drinking fountains and better lighting would be approved by management, giving the workers the idea that the process actually works and that they actually contributed toward making those decisions.

Deciding on a name for the circle is, in fact, the only completely autonomous decision quality circle members at UFC will make in the programme. The agreement between the union and management not to discuss collective bargaining issues also includes managerial prerogatives. Members explained that in addition to work rules and other union concerns, under no circumstances could managerial decisions be discussed. It was strictly off limits. As participants explained:

> There are certain taboos that you can't touch on. You can't question management about money or the way they run things.

> We were trained never to question management about some things. We're never supposed to tell them exactly what to do to solve a problem.

In order to foster a climate of cooperation and gain workers' trust and willingness to participate, management never clearly addresses the limited impact workers have in actually determining the outcomes of decisions made in the circles. Workers understand the restrictions on topics they are allowed to address. However, as we shall see, the constraints this places on choosing 'appropriate' topics and the actual *outcomes* of the decisions they arrive at begin to raise questions concerning the effectiveness of participation without control.

Establishing a framework for participation is a complex process. Quality circle members are encouraged to claim ownership of the process and use their experiences and imaginations to identify problems and provide solutions. However, at the same time, limitation on permissible areas they are allowed to address is extensive.

In order to best understand and analyze these processes, workers' negotiations of these limitations and the conflicts which erupted over them in the circles will be examined. By choosing this approach, we can locate sources of conflict and cooperation, examine the nature of the issues the circles address and the outcomes of their decisions.

Workers accepted at face-value management's claim that members 'owned' or, at least, had control over the direction of the circle. Members assumed that they would have the opportunity to choose appropriate issues and, once accepted by management, would be implemented. As we shall see, these assumptions were undermined by management's own projected needs for cost-savings (which workers did not have access to), by pressures from the quality circle leaders to tackle particular problems, and by the actions of managers who believed their authority was being undermined by the quality circle.

When workers talked about the projects they were initially interested in working on, they often stated that they would like to eliminate existing conditions on the floor that were a problem. Several discuss the issues that were important:

> We got together and decided, 'Let's tell them the way *we* would like it before they go ahead and do it, because they don't know what it's like to work here'. [an assembler whose work area was going to be re-organized]

> The main issue in our circle (CAD room) was how to get everything running more efficiently. There are so many copies of drawings floating around, you don't know which one is the original. It can get pretty confusing and out of control . . . We also have a lot of health and safety problems – lighting, lack of workspace . . .

> We have this terrible noise problem in our area. Everyone goes home with a headache. I thought maybe we could make it better for us, the workers. The company told us in the beginning that we were supposed to present ideas that would make our jobs easier.

> In our circle, we're fighting for a safe developer. The typesetter gives off an odor and we want to have it checked out.

These problems have little to do directly with cost-savings to the company (although they can be presented in terms of increased worker productivity) and instead reflect a determination by workers to address issues of concern to them. The groups' stated purpose is to solve problems in the work area and workers initially interpreted this in terms of their own needs.

The majority of issues raised by workers in the circles at UFC were health and safety related. They included: a spray booth in the shop, a room arrangement which emphasized safe operating procedures, excessive noise, chemical and solvent use, typesetter odour, and the transport of heavy cassettes. Out of the six proposals, two were approved – one was never implemented as requested – and the other followed only some of the proposal's recommendations.

Although a labour-management safety committee and elected union safety reps both exist at UFC, workers felt these issues had to be addressed in the circle. This is not surprising since OSHA, the federal agency responsible for safety and health regulations and inspections, has faced massive cutbacks in funding and the numbers of available site inspectors.[26] Even though OSHA continues to operate as the only means of legal appeal, unions have effectively lost the mechanism available for protection against unsafe or hazardous working conditions. It is, therefore, not surprising that circle members would consider health and safety issues as legitimate topics for discussion. As one of the members, who is also the health and safety representative remarked:

> We did it because it's extremely hard to get any kind of safety thing done at all with the company. With them it's, 'End of discussion. That's it.' It's like you can't get *anything* done.

However, in order for health and safety to be considered as a circle recommendation, members had to redefine its importance in terms of the rules of participation. Issues were to be analyzed in terms of cost savings, productivity and product quality.

The presentation by design drafters on a room arrangement for the use of the photo-plotter is an example of a redefinition of an issue to conform with the Quality Circle process. One of the members discussed the procedures for using the machine:

> It's basically a camera. We use it when we create a drawing on CAD and we want a print of it. After the machine photographs the drawing, we have to remove the cassette, which is quite large, and take it over to the dark room. It's a real hassle because we have to walk through a crowded corridor and the thing is pretty heavy and awkward to carry.

The final presentation the circle gave to the company focused on the 'inefficiencies' of handling the cassette. Complete with a booklet including photographs, drawings, models representing the different components of the problem and financial costs, the presentation emphasized both the health and safety hazards associated with the work along with the productivity and cost-savings per year of relocating the photo-plotter to a more appropriate site. Although members clearly addressed managerial concerns regarding cost effectiveness in the presentation, they prominently exhibited those factors contributing to unsafe working conditions. Members recommended that additional space be allocated to construct a dark room for the photo-plotter. They listed other possible solutions, but indicated all of the various drawbacks associated with them, including job jurisdictional disputes.

One of the members reflected on the process and its results:

> It took us well over a year before we were finally able to make a presentation to management. We worked on it very hard . . . When we got done with this presentation, we waited for this room to be built. Our proposal was to put this machine in a dark room which meant we wouldn't have to carry the cassette anymore – we could just carry the film.
>
> In our proposal, we indicated to management areas that would save them money – cost-effectiveness, time-effectiveness, cut down the schedules. We covered everything we could think of that would benefit them and benefit us.
>
> They agreed to it, but they didn't like our room design because, 'Well, we're gonna get new equipment in. Let's make provisions for this. Let's make provisions for that.' What we decided was a 12 x 12 room now become an entire redesign of the CAD/CAM area. It became quite frustrating. We had to let it go because management didn't really buy it. We more or less got a pat on the back, 'You did a great job. I'm impressed with all of you.' We heard all of that.
>
> It took over six months before they even started doing the construction . . . The point is, today the room is there. We gave them an alternative to modify the main computer room. They made several changes to benefit themselves. However, the room itself has not been completed. There are still four or five items that aren't done. I more or less got tired of asking in every quality circle meeting – 'What's the result?', 'What's gonna happen?' 'This isn't done, that isn't done.' I more or less became a pain in the ass.

This presentation provides a particularly interesting scenario of the nature and extent of cooperation and conflict within quality circles. Workers did pursue an issue which they considered a serious problem by incorporating it within the 'acceptable' Quality Circle format (that is, cost-benefit analysis). Their cooperation extended to the point of using the process to address their own needs. Conflicts arose when management ignored the very processes they encouraged workers to participate in, exposing, perhaps, their own agenda as well as a disregard for workers' efforts.

The feelings of involvement and the opportunity to exercise influence over decision-making were a part of the early stages of the quality circle process. Within the context of QC policy, workers did explore and address a number of issues and problems which affected their work and constructed changes and alterations in existing work practices.

The free expression and exchange of ideas associated with the process of brainstorming encourages the feeling of a sense of involvement and decision-making. Workers *do* participate in detailed discussions of production problems and their opinions on its probable causes and possible solutions are actively solicited within the boundaries of the QC process. As Kusterer[27] notes, this form of involvement allows management to take advantage of workers' desire to use the knowledge and experience they have accumulated on the job. One quality circle participant remarks:

> I think it's a way of management to know the value of employees by seeing the problems we tackle and seeing our results. It can get them to think, 'If these guys can really make a difference, they can't be dummies'. It also makes *you* feel like you have something to say about your job – that you're contributing to your job and that you made your job better.

The initial stages of the QC process usually involve intense debates and workers express a sense of individual and collective contribution to identify and solve problems at the point of production. Some members experience this as a form of workplace democracy, given that management encourages them to explore areas of production which are problematic. Some members recall this period as the most satisfying aspect of participating in quality circles. Juxtaposed to traditional bureaucratic mechanisms of decision-making, this process appears to be a source of empowerment. Moreover, workers initially believed that quality circles were a forum for *their* problems, adding to a sense of control over the initial phases of the process. As several commented:

> At first, everybody was enthusiastic – it was a lot of fun. We felt it could really help us out. You know, be good for *us*. Our first project took four months from start to finish. The floor plan was really detailed. I think it really shocked the company that we could put something together like that.

> I've been trying to get new members, to bring some fresh ideas into the circle. The circle won't survive unless we bring a whole new set of people in where they have enthusiasm that we had when we first started.

I've been involved in quality circles for about eight months. We had a project that took about four months and then we started another, but just dropped it. We just quit. The enthusiasm just dropped. Especially after we saw that nothin' was really gonna get done our way.

The initial feelings of success or accomplishment may be interpreted as a 'Hawthorne effect'. In some cases, workers were given what they thought was a relative degree of freedom in the choice of topics for discussion in identifying problem areas. (As we shall see later in the chapter, many workers eventually came to view this process as being management-directed.) Workers were participating in a new programme which, ideologically at least, claimed to be organized around issues of improving conditions (in order to unleash greater productivity) at the point of production.

Beyond the ideology of 'participation' promoted by management is the organizational structure which is developed to integrate workers. Regularly scheduled meetings during working hours, paid time away from the job, and management personnel who serve as advisors and represent top management at the meetings all contribute to a sense of importance to the project as well as the potential to take workers and their ideas seriously. That this process was instigated by management who, according to the company's Quality Circle propaganda, will institute changes recommended by workers, was a motivational factor to participate. It is the *opportunity* to possibly exercise influence over some decisions that generated a sense of enthusiasm among workers.

Nevertheless, the outcomes and implementation of their suggestions remain beyond their control. They effectively have no input into either the approval of their projects or their execution. Workers exercise no influence over quality circle decisions made outside of immediate circle meetings.

Conflicts between labour and management over what are to be acknowledged as 'acceptable' topics and/or recommendations characterize the Quality Circle programme at UFC.[28] In the circles of members who were interviewed, only three solutions proposed by members were actually implemented as suggested. All of them were from methods' circle and all addressed specific managerial issues of productivity and cost-savings. The circle in the sheet metal department received approval for new sets of tools for each worker and new rollaways (tool boxes). However, although management approved the purchase months earlier, workers were still waiting for the equipment to arrive. Moreover, the agreement reached in sheet metal was a compromise. It became a package deal:

management agreed to the new tools as long as the uniform tool boxes were accepted.

Any problems associated with health and safety were either resisted or outright rejected in circle meetings as 'inappropriate' topics. Members complained that management often ignored their suggestions or downplayed them when choosing among topics for discussion.

The definition of a 'suitable' topic is part of the struggle for control over quality circles at the point of production. Workers joined the circle to further their own interests and exert influence over working conditions. Rather than seeing workers as simply reacting to managerial programs and the ideology behind Quality Circles, these data show that members bring their own agenda to circle meetings. For them, interpreting quality circles in terms of their own needs is not inconsistent with the stated aims of the programme – increased job satisfaction and the improvement in *their* working conditions.

QUALITY CIRCLES AND LABOUR-MANAGEMENT CONFLICT

The deliberate ambiguity which management fosters in defining 'participation in decision-making' lends itself to creating conflicts over just what the term does represent. Those struggles are evident at different stages of the Quality Circle process when the concerns of labour and those of management collide, laying bare the conflicting interests each group has in the programme.

Two major areas of conflict are: first, those within the Quality Circle process itself: the manipulation of the process toward addressing management issues, management's resistance to address and approve health and safety proposals, feelings of cooptation resulting from perceived manipulation, and misrepresentation of quality circle discussions and activities in the minutes of the meetings. Second, management's responses to proposals and decisions made in the circles: no action taken once proposal was approved, the reorganization of the presentation to fit management's agenda, sabotage by lower level managers, and approval by lower management and supervisors, but voted down by upper management.

Managerial sabotage of QCs is well documented.[29] Inefficiencies in production, worker morale and product quality have, traditionally, been the responsibilities of supervisors and local managers. Their resistance to quality circles stems from their fear of exposure as inept leaders to upper management, as well as their unwillingness to relinquish any control over traditional managerial decisions to workers. One member complained:

The layout we did for our area was really well planned. We left enough room between machines for safety precautions. The floor supervisor worked with us to figure out just how far these machines should be set apart. Then we get this other guy – I guess he's the planner of plant layouts – he came down and said, 'This is no good', and changed everything around. We said, 'Wait a minute. We got this agreed on. We got this on paper'. The thing that bothered us was that he didn't offer any explanation. He just said, 'This is the way it's gonna be'.

He was the one who originally combined all the assembly areas together and I guess he felt his toes were stepped on. They were putting drill presses right next to one another. There's one drill press that's totally useless because it's right next to an entrance. It's like he did it because we stepped on his toes.

But when the millwrights came in and moved everything around, we were telling them where to put them. We work with the millwrights a lot, so we were able to, you know, work it out with them.

Two organizational factors affect approval of quality circle decisions: traditional lines of authority remain intact and are not significantly altered by quality circles, and the quality circle steering committee – as an additional formalized decision-making body – has the authority to set policies concerning all activities related to the Quality Circle programme. The non-bureaucratic, informal style of the meetings, including the emphasis on worker-identified problems and solutions masks the actual structure within which quality circles operates. Once the problem-solving processes and solutions are generated, workers no longer have influence over the outcome of their suggestions. It is a process Ramsey[30] aptly terms, 'phantom participation'. Participation and decisions by workers are relegated to the immediate circle leaving the power to decide and act on their recommendations within the traditional bureaucracy. As one member put it:

In reality, it's the company that really makes the decisions. We're just sitting there and being told to come up with a few ideas and if we [company] could do it your way, we might try it. But if not, we're gonna do it our way, anyway. A lot of the guys felt that they were just getting a bunch of good ideas. They felt that they were just being used.

Because quality circles do not replace the bureaucratic hierarchy, nor do they radically alter the division of labour between labour and management, managers continue to be responsible for maintaining control within their respective departments. Conflict over 'domain' emerges when circle recommendations infringe on decisions usually made by management. Supervisors and middle managers usually have little or no input into approving projects. That decision is made by upper management. Several members complained about management's recalcitrant behaviour toward quality circle decisions:

> It's management – basically our immediate supervisors – who are the ones to hurt the circle. If you come up with something that happens to be in their territory, they'll try to fix it before the circle gets to it to show that the circle isn't needed.
>
> We had a few problems that were accepted by management, but they sat on it and never did anything about it. And that was nearly two years ago.

Quality Circle propaganda claims that traditional supervisory work is not – and should not – be adversely affected by discussions and decisions made within the circle. The stated aim of the programme is to promote ideas for productivity improvements without reacting punitively toward managers who appear responsible for that work. Nevertheless, as several members recall, supervisors and lower level managers are often held accountable:

> This guy from management came down to a meeting. He heard us talking about this tool that wasn't calibrated. He didn't say nothin' there at the meeting. He just sat there taking notes. When he left, he made some phone calls to our department and there was a big argument.
>
> The problem [with quality circles] is that you're sitting there with management who are the ones doing things wrong in the first place. The very people you're criticizing are there – and they don't forget it.
>
> A lot of guys in my circle felt intimidated. The leader was our supervisor who acted like a supervisor in the meeting. I think they [supervisors] feel threatened and feel like they have to be in command.

A report in *Iron Age*[31] discusses the pivotal role of supervisors in successful QC programmes. It notes that the immediate supervisor can be either the strongest or weakest link in the process. The foreman's role is changing from task master to communicator between labour and top management.

Managers who did participate as Quality Circle leaders often used the opportunity to encourage workers to address problems for which supervision was responsible. Under ordinary circumstances, they would have to solicit workers' cooperation once plans had been made. Here was a situation where workers could come up with solutions to these problems which they would have a stake in implementing, thus reducing, if not eliminating, their resistance to management's plans for change.

Moreover, this situation could provide management with access to workers' extensive knowledge of shopfloor practices. Wells[32] found this to be the case in his study of Quality Circles. Management often used mechanisms set up to enhance productivity as a means to gain valuable information on working practices. At UFC, information was solicited through the 'brainstorming' sessions which characterize quality circle meetings. If workers accept a management-defined problem as an appropriate quality circle issue, the process of identifying all potential causes might indeed include their knowledge of work group practices.

In addition to the information solicited, the processes associated with finding solutions and gaining management's approval are not subject to the same resistance and restrictions that most worker-oriented suggestions face in the circle. The approval rate of methods' proposals is an example of management's role in quality circles and their ability to elicit workers' 'know how' of production processes to improve productivity. One member explained:

> Yeah, we did have some input from management. They tried to give us some kinds of ideas of what projects to work on. They'd say, 'We're having a problem with this (feasibility of doing castings against solid pieces). We'd like you to look through it and see what you can do to perhaps solve it.'
>
> One of them – where we would figure out the deductions and allowances on the computer, rather than have the inspectors use the calculator. Management accepted it even before we made the presentation. That happened to be the big manager's pet peeve and when he heard about it, he said, 'Good. We'll use it'. He spoke to the supervisors and asked them to accept it and do it the way we said to have it done. That was like a feather in our cap.

Other members, however, resented and resisted management's attempts to dictate (whether directly or subtlely) the direction of the sessions. The loose structure of Quality Circles enables managers to use different types of coercive strategies to impose their ideas in the Quality Circle process, including showing a visible lack of enthusiasm for ideas, persuasive discourse, and/or active intervention to influence the direction of the discussion. Some strategies have been apparent, while others are more carefully concealed by verbally camouflaging intent and/or creating conditions for workers' acceptance of them. Several members commented on the ways management manipulated the process:

I can tell you one thing I've noticed – and you can be sure of it – a supervisor who is a leader and manager of a department will use quality circles to their advantage and lean on it if they have to. If the circle gets into union issue and you approach the company on this, they say, 'Well, the quality circle decided to do this, and they should be autonomous. We don't want to interfere.'

You can bet that if it was something that management didn't want, they'd interfere right away. And I've seen that happen. A circle decided on a certain set of procedures to follow in making changes in the way things were done. Management disagreed with at least two of them and said, 'No, I don't care what the quality circle says, we're not doing this'. The circle was pretty miffed about it and a few people dropped out, but no pressure was put on the manager.

I don't think they [management] let the circle go the way it wants to go. We have a sheet metal supervisor who is the circle leader. Unfortunately, a lot of people are easily swayed about what should be done . . . And he's in a good position to do that . . . He doesn't come out and say, 'Well, I'm the boss and I think this idea stinks'. He'll sort of lead people.

Like the thing with re-arranging the shop. It was one idea that came up. He pushed for that without coming out and saying it directly. He said, 'In my opinion, it's probably the biggest problem on the list'. Whether it is or not might be two different stories. It's easy to direct people. They can be easily influenced.

There were a lot of flare-ups in the circle. The supervisor (also a quality circle leader) is not the easiest in the world to get along with. A lot of people dropped out because they didn't like going down there and getting yelled at for an hour. We called it 'vicious circle'. All he wanted us to do was to work on increasing productivity. That was *the* main thing – productivity increases.

In many instances, management attempted to continue to exercise authority within the circle, even though, according to the rules of the pro- gramme, traditional forms of decision-making were eliminated, replaced by mutual trust and sharing. Workers did complain that they were being manipulated toward addressing issues they believed were clearly management's problems – tape storage, passwords for the computer system, room arrangements and various ways for workers to 'manage' their own productivity.

Workers resisted many of these attempts. In addition to setting their own priorities (which were often ignored and/or downplayed by the quality circle leader) members viewed management's behaviour as attempts to monopolize what they considered their domain or, at the very least, a kind of 'neutral territory'.

The most blatant example of attempts at controlling and manipulating the circles was the publication of the minutes of the quality circle meetings. For members, the minutes were, in fact, a distortion of the ideas, issues and comments raised in the meetings, as well as a conscious attempt to eliminate any suggestions of conflict between labour and management and dissension within the circle.

At every session, a circle member was responsible for recording the minutes. The quality circle coordinator would have them typed and distributed to both quality circle members and the steering committee. According to members, the final versions of the minutes differed markedly in both content and tone from the original minutes and actual events. The extent to which traditional labour-management conflict becomes a part of the quality circle process is evident in the following quotes:

> Every week somebody else took the minutes of the meeting. They would write down exactly what happened, exactly what the group said about certain things (like our supervisor holding people back from the meeting). We would give it to our coordinator and she wrote in in different wording, and left things out almost 100% of the time. There were points we wanted to get across because we knew copies were being sent upstairs. And we were told when we first started that we were the voice to the company. How are we going to be the voice to the company when they're going to change the minutes and the real facts?
>
> The coordinator says, 'Well, you can't make waves'. We said, 'Listen, that's what we're here for'. A lot of guys dropped out 'cause they figured it was useless. We're made and we want to tell them why. We were told from the very beginning, this (QC) was our way of letting the company know how we feel about certain issues. It would be beneficial to them, not only us. But everything we did was beneficial to them.

In my minutes, I put such and such chemical could possibly cause cancer and they wouldn't let me put that. Their point is 'Don't stress the health hazard. Stress that you can't get the job out.'

They're trying to make us believe that if we go from that point, management will listen. In other words, getting the job out is more important than if we drop dead of cancer. I say, 'You want proof, we'll get the proof, but untie our hands'. I think they're trying to get us to believe it should go their way.

The biggest issue now is who's gonna write on the blackboard and take the minutes. They didn't like what I had to say. They said I was too direct. They make you change the wording. I could say, 'We're getting nowhere and management is deaf, dumb and blind'. Or, I could say, 'There are some difference between us and maybe with a little persuasion management will listen to what we have to say'.

I would run it where there's freedom of speech. Don't sit there and dilly-dally – 'Well, let's see the *best* wording. You know in your heart what's right. Don't spend fifteen weeks stating the best way to phrase it.

My feeling is – take hostages [she laughs]. I feel we should be an independent group.

The tensions and disagreements which erupt in the circle reflect both traditional bargaining issues (health and safety) as well as managerial attempts to determine the nature and outcomes of workers' participation in quality circles. The minutes of the meeting represent, in concrete terms, the social relations unfolding within quality circles. The notion of democratic process is denied as management reconstructs events to conform to a more ideologically compatible image of the process as cooperative rather than adversarial. The propaganda of cooperation and participation is doubly contradicted here. By this action, management distorts and denies the issues workers believe to be legitimate concerns, and attempts to defuse and depoliticize the context in which workers define issues and propose solutions to these problems.

Without question, any documentation of dissent would reflect on the QC coordinator's ability to successfully operate the sessions and create a climate of cooperation. The coordinator, as well as management in the steering committee, have a stake in presenting QC labour-management relations as non-adversarial and productive. Their effectiveness, as managers of an experimental programme, depends, in part, on their ability to 'package' Quality Circles within the rhetoric of cooperative labour relations. One member, frustrated with the lack of progress, commented on management's reactions to workers' proposals:

I told this guy **** [in charge of UFC's QC programme], 'You've got all this wonderful stuff in your newsletter. I never see any of this. What's happened with all these projects? Where are the results? Show me *anything* that's positive about QC.'

I wanted to see what they've actually solved and what's happened with it. Those newsletters are really all candy-coated to make **** and the group under him look like they're doing a wonderful job. They're not going to tell anyone there's anything wrong or there's really nothing going on. I think that's what's going on with the power structure of quality circles. I said to him, 'You print all this stuff and it's all lies!'[33]

Beyond the issue of the importance of management's successes with the programme remains the larger question of workers' role in the process. Any claim to ideas and discussions in the meetings, represented in the form of minutes, is challenged by management's control over its final version and distribution. Workers found their ideas consistently subjected to review and reinterpretation. They are marginalized from a process which depicts their actions in the circle, revealing, in effect, the inequality in the QC process. It is more than an example of management's need to present a 'positive image' of their operation of the programme; but of conflicting interests between labour and management around the meaning of participation.

The lack of workers' power and authority within the circle is made obvious by this action. Management's response to disagreement is to deny its existence and replace it with a version of events that reflect their own goals and interests. Not surprisingly, workers resented and were angered by this move. The minutes, unlike much of the interaction which occurs during brainstorming or problem-solving sessions, are tangible evidence of conflict between labour and management over issues and strategies. The generally less authoritative approach of circle meetings often camouflages disagreements by its emphasis on compromise and consensus. However, this context is challenged when issues and events are concretized in written form.

QUALITY CIRCLES AND UNION ISSUES

Less obvious than the conflict over the accuracy of quality circle minutes – but no less problematic – is the infringement of the circles on union issues. Explicitly forbidden by the terms of the agreement between the union and management regarding participation, the evidence, however, suggests something different. When asked directly about the likelihood that the circle might inadvertently address contractual issues, workers denied the possibility. All claimed to be well aware of the importance of avoiding discussions of the contract and saw the programme as fundamentally different in its concerns from the union agreement.

Moreover, shop stewards were to attend the weekly meetings to ensure compliance with this rule. Since workers had approached the project with their own sets of issues – and believed they were acting in their own best interests – they saw no reason to be overly concerned about jeopardizing union solidarity. Given these conditions, workers were confident that they would not undermine the contract.

> I think the union members who are in the circle should know the guidelines of what is union and management issues. They should keep it in mind at all times in the discussion. Each individual should look in the contract and know what is out of bounds.

> Right now they're (stewards) there to represent the union to make sure we don't talk about things that are in the contract. I think people in the meetings are intelligent enough not to talk about union business . . . I think in the beginning they wanted to make sure it was run the right way. I don't think they should be involved at the level of the circle, only at the upper levels with planning quality circles.

> Things would come up about contract issues. But everybody would jump in and say, 'Can't talk about that. That's in our contract!'

Comments like these were repeated in conversations with quality circle members. There is no doubt that, aware of infringements, workers would defend their contract. Nevertheless, structural factors inherent in the quality circle process, along with the particular dynamics of the group's decision-making apparatus, create opportunities for intrusion into the area of collective bargaining.

To understand how the process operates to undermine union strength on the shopfloor involves taking a closer look at both Quality Circle structure and the dynamics of the circle. For this, we need to briefly return to an examination of methods used to identify factors affecting productivity – chief among them, production methods and manpower.

As discussed earlier, workers are trained to identify those processes which contribute to solutions for increased efficiency and productivity.

Two of these areas – methods and manpower – represent complex issues around the definition and use of skills, which have historically been part of the collective bargaining process. As key factors in QC decision-making, it is impossible to carve out those aspects of the labour process which are not covered by the contract.

Moreover, complications arise since what constitutes bargaining issues is always a point to be negotiated both at the bargaining table and at the point of production. To limit *potential* areas for collective bargaining and/or infringe on existing contractual agreements – particularly those which are susceptible to interpretation – weakens the union's ability to protect workers in present and future bargaining sessions. Because QCs focus on productivity, they do, in fact, address contractual issues.

This is directly at odds with the way in which the union perceives its role. As one former union official (also on the QC steering committee) said: 'Anything and everything that bothers an employee that management won't address is a union problem.' However, by agreeing to cooperate with management around issues which are a source (or potential source) of conflict, unions may weaken their position as workers' representatives, particularly if they adopt the 'win-win' strategy promoted by QCs. In order to bridge the gap of 'adversarial relations' which reflect traditional labour relations, common ground needs to be identified and issues formulated around 'mutual benefits'. The notion of a 'win-win' situation captures the spirit of QC programmes and focuses on compromise as a way around issue of conflicting interests. The emphasis is on identifying issues and solutions where labour *and* management can both be winners.

The circle in sheet metal provides an example of how this strategy works. One of the shop stewards commented that sheet metal 'got a quality circle because management thought there was a "morale problem" in the department'. The facilitator then encouraged the circle to adopt uniform rollaways in exchange for each worker receiving a new set of tools. In the past, sheet metal workers used their rollaways as a form of individual expression.

> We're all supposed to be getting these uniform rollaways. All it is a tool box on wheels. Some guys build on theirs – they look like three-story condominiums. I guess it looks better for the company. No strange bumper stickers . . .
>
> I don't go for them too much. I like the personality to the tool box – to the eight hour day. If we get the rollaways, all that has to go – no bumper stickers, no pictures, no personal items . . . I really don't go for that. I'm into freedom.

This agreement has two implications for the shop. First, it is a trade-off for sheet metal workers. In exchange for receiving new sets of tools (which will increase productivity), it is also an agreement to restrict personal expression on the floor and a sense of control over their spatial environment. Second, it removes work from another occupation in the bargaining unit. As one shop steward remarked:

> The idea of everyone getting their own set of tools sounds great since you have everything there for you. It's definitely more efficient for the company. But we have a man who hands out the tools and he says, 'It's my job. What am I gonna do next?'

The opportunity to assess the impact of decisions made in quality circles on workers in other areas in the plant is limited by the design of QC around specific productivity issues. Nevertheless, within the QC framework of 'non-adversarial' relations, such an agreement is viewed as a 'win-win' situation, with workers and management gaining some advantage from the outcome. The notion of 'winning', however, is linked with the goals of the QC programme – improved efficiency and productivity. The implications of what is actually won (and lost) is never clearly articulated. Without fully understanding the meaning of this strategy for the organization of work, workers agree to changes they would normally resist in traditional collective bargaining agreements, such as the removal of work from the bargaining unit.

The dynamics of the sessions – brainstorming and proposing solutions – expands the potential areas the group is able to address to include virtually every aspect of the labour process. Within the context of 'open relations' and 'cooperation', this process creates an opening into union issues. Workers adopt a less guarded approach to problems and solutions when given responsibility to identify and solve work area problems. The particular interactive nature of the QC process encourages this response.

This is most noticeable when the chosen problem involves other occupations at a different stage in the process. A member explains the impact of one of their recommendations on keypunchers:

> Our first presentation was involved with reducing the time that the OSs [operation sheets] spend in the keypunch room . . . At one time, if you changed one number on the OS, the whole thing had to be sent down to keypunch all over again. We said, 'Why don't we eliminate that by having a form delivered with it telling the people what was changed and just that portion would get changed.

The service section – the clerks – just sent down the pages that were changed. So that reduced their work. Something was said about that – that we probably eliminated some keypunching jobs. Actually, I don't know if anyone was laid off because of that . . . We saved something like 500 000 hours. It came out to a savings of one million dollars, so it was quite a substantial savings to the company.

Workers participating in the programme are caught in a complex arrangement which – although it denies the existence of labour-management conflict within the process – also embody traditional struggles over controlling the labour process. In addition, the locals, after agreeing to be 'partners' in the programme, maintain an uneasy and difficult role in supporting members' participation and intervening when management *and* workers begin to address collective bargaining issues.

A former union official discusses a problem he faced when a circle began a project which would have directly undermined the contract:

The issues raised in the circles really varied. Some were very practical and others were less tangible. Sometimes union problems got involved in there . . . In the quality circle in the drafting department, they wanted to change the process in which draftsmen learned to do their job. What they wanted to do was to give design-type work to drafting people. They wanted to give a higher (technically) assignment to a lower paid classification so they could learn as they're going along.

On its surface, it's great. However, the only way we've ever been getting any movement between these occupations was when you noticed that supervision was trying to get the most work for the least amount of pay. It was only through grievances that we filed which would make openings for design people; otherwise, there would *never* be openings and people would never move.

Where, in the circle, it looked like an innocent type of thing. It had a lot of sincerity behind it; but it ruined certain things that the union was trying to do. At the time, we [union] were talking about having a type of automatic progression instead of the test system. At the same time, they would be able to give any work they wanted to anybody. That's what we were planning to give up, so to speak. If they already had that, they could assign work to anybody they wanted and have absolutely no reason to go along with this automatic progression. It causes a problem.

At that point, I had it stopped. Even though these people mean well, we can't have it. The area rep resented it and I guess a few members did also. I told him, 'If there's any problem with it, tell them [members] to come and see me and I would be glad to explain it to them'.

There are several important issues raised here. First, the plan reflects the concerns expressed by design drafters and drafters regarding the effects of CAD use for the acquisition of design skills. Fears of an occupation in decline and blocked chances at upward mobility within the occupation are critical issues for these workers. Second, as quality circle members they saw the possibility of preventing this loss by creating a system for acquiring knowledge. Third, the insistence by members that 'no union issues are discussed at meetings' appears valid to them since the workers identified the problem and worked on a solution they believed would benefit themselves and the occupation.

Moreover, CAD training is examined in a context which emphasizes labour management cooperation. The 'win-win strategy' approach masks the real threat to the union's ability to effectively represent its members. The plan recommended by the circle would essentially make the grievance procedure irrelevant and significantly reduce the power of the union to protect members. Far from enhancing workers' control over the acquisition and use of skills, the formulation of these issues through the QC process by workers directly involved serves to weaken the union's ability to define issues and responses to change in the labour process through collective bargaining. Its support for the programme places the union in a difficult position of criticizing its members' ideas while continuing to participate in the QC process.

Other circles have resisted managerial attempts to routinize and simplify the labour process. Managements' suggestions to circle members for the creation of step-by-step procedures for setting up CAD work were rejected outright by members as managerial issues. Further, workers viewed this as a threat to the control over their skills. Because CAD is still a relatively new technology, designers working on the system have virtually complete control over the process.

Wells identified this kind of QWL decision-making as part of a new management control strategy. Along with the delegation of a certain amount of authority to workers in order to police themselves, Wells cites 'improved access to workers' skills and knowledge about their jobs'[34] as a key component of this strategy. Under ordinary circumstances, workers guard the formal and informal skills and knowledge they have accumulated. Management's access to this critical information provides an opportunity to reorganize the division of labour toward simplification, fragmentation and routinization to the extent that it is feasible. The development of systematic procedures would remove the autonomy of designers and give management a central role in determining the use of skills.

The effects of such a strategy can lead to a weakening of solidarity among union members. Discussions of cost-savings can involve members identifying practices within the work group – along with particular workers – which affect productivity. Moreover, this process may result in workers identifying with and supporting suggestions made to increase productivity. One steward complained about its effects on the union:

> They talk about trying to get the job done quicker and easier. There are guys in there who feel they *have* to tell them because they agreed to be involved . . . Sometimes it gets into some finger-pointing . . . Now, they're talking about the problem of going down to the 'oil room' to pick up the epoxies we work with. They're saying that they should hire another guy to go down there and get it for us, so that we don't have to leave our job to get it. We could still be doing our work while the guy is getting it for us. I say, 'Hey, if they want to pay me to go down and get the stuff, I'll go down there and wait on line for an hour and come back'. I enjoy the time away from my job. I get to talk to some of the guys I wouldn't be able to do otherwise.

Workers historically have used job responsibilities (for instance, going to other departments for needed supplies, information, and so on) for an unofficial break and the opportunity to socialize. It is often a way to control the monotony of a job and the pace of a workday. This movement also maintains informal communications between groups of workers to discuss shopfloor problems and grievances. Wells views workers' responsibility for productivity and limited authority within the circles as undermining 'the main form of power that workers and their unions normally resort to – the negative power of resistance or refusal to obey'.[35]

The ability to resist and the nature of resistance is affected by the ways in which issues are defined and the mobilization of resources (including workers' ideas) are utilized. According to Wells,[36] participation in QWL-type programmes often weakens the kinds of links workers establish to effectively deal with conflicts with management. The possibility of cooptation exists particularly when productivity issues are the major focus of group discussion. There is the danger that workers will begin to 'second guess' their abilities to develop appropriate strategies in defence of their own interests, particularly since the QC process attempts to gain workers' understanding of the problems facing management in the attempt to remain competitive in the market.

WORKERS' RESISTANCE AND THE UNION

Despite management's success in generating productivity and cost-saving suggestions from the QC process, workers did resist many of these efforts and fought to have their own issues addressed. Given the locals' partnership agreement with management, how was the union able to respond to workers' dissension over the programme and yet continue to remain management's 'non-adversarial' partner?

As stated earlier, the union agreed to participate in the programme only after certain demands were met: no contractual issues were to be discussed, and the right of union representatives to be present at meetings to ensure against violation of this agreement. With these assurances, the union then agreed to become a 'co-partner' in the QC steering committee to plan and manage the programme.

The 'top down' structure of this policy-making committee effectively removed the union from direct involvement in day-to-day circle activities. Since the shop stewards only monitor infringements on the collective bargaining agreement, there is no liaison to the union to report members' views of and experiences with the QC process.

Moreover, the progress reports and data provided to the steering committee are generated by management. The decisions in which union leaders participate are made around management's specific concerns for productivity, efficiency and cost-savings. In essence, the role of the steering committee is to set an agenda for the QC programme which reflects these issues and monitors the progress of circles in relation to them. The union's role is limited to guarding basic contractual items from infringement by the circles and maintaining labour's participation in the process according to agreed-upon managerial goals.

Some union leaders viewed participation as an opportunity to make gains on certain issues which they have been unable to achieve in collective bargaining. They believed that through the QC 'win-win' strategy management could be convinced to agree to changes by emphasizing their 'mutual benefits'.

This strategy, however, is flawed. Since the focus of the bargaining strategy is altered to reflect concern for *management's* productivity needs, it has potentially negative effects on union solidarity. Protecting workers' rights becomes obscured by the QC context of compromise and emphasis on productivity improvement. And, as Wells[37] points out, the QC process itself has a tendency to fragment collective interests into discrete problems, further jeopardizing the union's strategy.

Similarly, the agreement to adopt a 'non-adversarial' approach toward participation left the leadership unprepared to deal with conflicts that emerged in the circles. As the data demonstrate, management undermined workers' ideas and sought to control the process and outcomes of ideas generated and decisions reached in the circles. In one case, workers' collective angry response to this manipulation caught the union by surprise and received a confusing reaction from the leadership. According to the shop steward:

There was an uproar today. The stuff hit the fan. All the members want to quit. They're worried because the contract is coming up and they might lose their jobs by giving them all this information . . .

Everybody's saying 'Why should we be telling them how to make our jobs easier? That's for us to know. Let them think we're breakin' our bananas about something.'

So **** [union secretary/treasurer] comes in this morning to talk with us. He said the union in behind it 100%, but that we should do what we want. It left these guys saying, 'Did we do wrong by saying this?' He left them hangin'. So, now the guys are stuck in the middle saying, 'Well, if the union is behind them [QCs], how can we say it's gonna affect us?'

We don't want to be any part of it because as union members we think they're wrong; but here comes our secretary/treasurer who says 'We're behind them 100%!' We wanted to show them [management] union power and show our support for the union by saying, 'Forget these quality circles – we don't want no part of it!' As the shop steward I told them, 'If you guys think you're right, then you're right. Don't let anybody tell you different'.

The union's acceptance of a limited partnership in the operation of QC and of the 'non-adversarial' relations which (ideologically, at least) define participation left them unprepared to deal with the conflicts which arose within the circles. The leadership believed, erroneously, that given constraints on discussing collective bargaining issues in the circles, and their positions on the steering committee, they could maintain a successful separation between traditional labour relations and QC cooperation.

Operating under these assumptions, they failed to consider the possibility that union strength could be undermined. When confronted with a revolt by members, the leadership responded by treating it as a *QC* problem, hence the advice to quit if they were dissatisfied. In effect, the union became a mediator of QC conflict and acted to ensure its members' continued cooperation. Workers found themselves caught between resisting what they believed was the cooptation of the union, endangering their contract and the informal exercise of control at the point of production, and the union's assurances that participation would not affect union solidarity nor collective bargaining. The leadership's actions served to isolate workers' dissatisfaction and removed critical union support for rank and file dissent.

Disagreements between labour and management in the circles over the control of the process led workers to challenge both the ideology of QC as democratic participation as well as the actual impact such cooperation was having on workers and on union solidarity. The high dropout and turnover rate of participation in QCs are evidence of worker dissatisfaction with the programme. However, because it is voluntary, workers are free to leave at any point. Resignation is treated as an individual decision rather than a reflection of the undermining of the collective interests of workers. The decision of members to act together through a group resignation challenged both the motives and 'non-adversarial' nature of QCs. The union, believing it had safeguards to protect workers' interests, was caught virtually by surprise by the level of fear and resentment the programme had generated.

This points to a fundamental problem facing unions who agree to participate in and support managerially instituted schemes such as quality circles. The very structure of QC – the bureaucratic format of the steering committee and the QC process on the floor – subsumes union issues within it. In most cases, the leadership is called in once the major decisions on and primary focus of the programme are already determined. They are, then, consulted on the *details* associated with gaining their members' cooperation and with contributing to the implementation of the programme.

Unions, therefore, become unwitting partners in supporting not just productivity improvements, but a weakening of the collective strength of its membership and its ability to assume a leadership role in conflicts which do erupt. However, although workers' actions, such as those stated above, indicate the difficulties and dangers of participation, they also point to the resistance to policies and practices designed to undermine their control on the shopfloor.

SUMMARY

The data presented show that Quality Circles are implemented primarily as mechanisms to enhance productivity, efficiency and cost-savings. Benefits to workers are secondary at best and participation in QCs does not lead to any significant role in decision-making or increased workplace democracy.

The structure and process of QCs provide evidence of the ways in which workers are included in the programme at the same time their participation is circumscribed to conform to the specific productivity needs of the firm. As the data demonstrate, QCs are not autonomous from traditional bureaucratic organization. Although QCs are somewhat horizontal arrangement, they remain part of the formal organization of a corporation. This organizational structure is an example of the movement toward decentralized decision-making around issues of productivity and efficiency within a plant and at the point of production.

The reorganization of corporate planning into committees coordinating operations at the executive level, the decentralization of decisions regarding local operations, and the implementation of a QC programme that facilitates local management's decisions around productivity are all part of UFC's restructuring process. The role of the QC steering committee is to operationalize and monitor that aspect of decentralization which includes workers' involvement. The steering committee is, in fact, part of *management's* restructuring around issues of productivity and efficiency. It is in this committee, which includes union leadership, that the ground rules for participation are decided. However, the cooperation of the union in the functioning of the programme creates contradictions for their position as workers' representatives.

The union was an unequal partner on the steering committee. All reports and data were generated by and for management's own requirements for the programme's success. The minutes of the meetings – the only systematic written information the union had on its members' participation – were re-written by management to reflect harmonious and cooperative relations within the circles.

The leadership found itself at odds with its membership over decisions reached in the circles which would adversely affect collective bargaining and traditional union protections. The separation of QC and union issues ultimately was not as clear cut as workers and the unions claimed it would be. The presence of the shop steward in the meetings was not a guarantee that union issues would be avoided. Furthermore, the techniques associated with decision-making – most notable, brainstorming and problem-solving

– are designed to push the limits of these boundaries into virtually every aspect of the labour process.

The nature of labour-management relations within QCs, despite the veneer of cooperation and equal partnership, were, in fact, part of traditional labour relations. The dangers to workers – and to union strength – are greatest where the accumulated knowledge of production methods, skills and informal shopfloor arrangements become a part of open discussions with management on the means to increase productivity. In this case, management's response was to use what information they did gain to further their own needs in reorganizing production, fostering suspicion and resistance among workers who participated in the process. Disagreements with management over appropriate QC issues, control over the publication of the proceedings of the meetings and over the outcomes of workers' recommendations are indicative of the ongoing conflict to exert control over the labour process.

Ultimately, the project was a failure. Contradictions between the ideology of cooperation and the actual purpose and outcome of QCs led to a high turnover rate among participants dissatisfied with the programme. More seriously, these contradictions produced confusion and uncertainty among workers who viewed QCs as detrimental to labour solidarity. The support of the union for the programme in the face of this discontent only served to add to workers' sense of insecurity about the meaning of their participation.

After the interviews for this work were concluded, two of the locals withdrew their support for the QC programme. As contract negotiations broke down and workers went on strike, the remaining locals ended their participation, citing management's shallow commitment to cooperative labour relations. QCs are a continuation of management's historical efforts to redefine and control the labour process as well as a contemporary trend in corporate restructuring.

The union's initial agreement to cooperate with management in the QC programme as a means of gaining job security and dignity for its members is not without potential costs to the union's leadership role among its members and harm to its own collective bargaining strategies. Nevertheless, as the data show, the resistance of workers to participate in a process they come to view as manipulative and deceptive is an indication that they will continue to act in their own interests.

The idea that workers create significant changes through QC participation without jeopardizing the informal and formal mechanisms of control which they have developed at the point of production and through collective bargaining is not supported by these findings. At best, workers were able to make small gains around specific working conditions. However, the potential for undermining labour solidarity and union strength existed throughout the programme at UFC and remained a threat to the collective bargaining process.

Nevertheless, workers resisted management's strategies to control many of the processes QC propaganda claimed were to be workers' decisions. In addition, the lack of any real control over the outcomes of their decision-making only increased workers' resentment. Ultimately and inevitably, it was conflicts over the issue of control which led to the decline of workers' support for the QC program at UFC.

11 Conclusions

This book set out to study several changes which have recently been or are presently being implemented in industrial organizations and labour relations. The analysis has yielded several key findings. First, the impact of computerized automation on the labour process is complex, contradictory and multi-dimensional. The comparable effects on the skills of machinists, methods, drafters and design drafters are not uniform, nor can it be simply categorized as de-skilling or upgrading within each occupation. Moreover, the findings in this study clearly show that automation is not the singular 'cause' of changes in job skills and the organization of production. Managerial decisions about the definition and use of skills affect the implementation of automated equipment to the production process.

Second, Quality Circles, rather than being a transformation of labour relations toward democratic participation are managerial strategies to create a decentralized organizational structure to facilitate workers' cooperation with management's goals. Management's propaganda to encourage participation focused on ideals of worker dignity and workplace control. The contradictions between the rhetoric of participation and the actual processes and outcomes of the programme ultimately resulted in its failure. The narrow range of 'acceptable' QC topics, disregard of workers' suggestions for change, and the introduction of managerial concerns into the QC process for productivity, efficiency and cooperation contributed to workers' discontent with, and withdrawal from, the programme.

Third, the transformation of industrial relations goes beyond the schemes of participation and decision-making symbolized by QC and other QWL programmes. The movement toward the flexible use of skills through the merger of job classifications and the elimination of traditional demarcations between occupations are tactics to extend managerial control over the labour process as the workplace is reorganized. Although not a direct outcome of computerized automation, these processes, nevertheless, are affected by the flexibility of the technology.

The findings clearly indicate that the impact of computerization on the labour process is contingent upon the social relations within the firm. It is this arena which is the 'contested terrain' between labour and capital over the control of the labour process. And it is here that the technical reorganization of work assumes its particular form. Technology both influences

200

and is influenced by social factors, including the current level of workers' skills, the nature of the industry and its product markets, unionization and the history of labour relations. Social relations are conditioned by the possibilities and constraints of the technology. They also simultaneously shape its use, define the reorganization of the labour process and the extent to which work is de-skilled and/or re-skilled, the content of jobs is altered and the hierarchical structure of occupations reconstructed.

As the data clearly show, the use of skills and the control over them in the production process is shaped by both organizational decisions and labour-management conflict. UFC's production of complex and special-ized parts, small batch production methods, highly skilled and unionized labour force, existing divisions of labour, worker resistance and collective bargaining agreements all contribute to the decisions surrounding the use of automation within the plant.

Computerized automated work is composed of two distinct, but inter-related processes: traditional skills which have historically defined a craft or occupation; and abstract cognitive processes associated with processing computerized information. This analysis takes both processes into account. The lines of demarcation which have historically separated occupations were established primarily by task differentiation. The critical importance of information in an increasingly automated workplace transforms specific tasks into a generalized application of knowledge about the entire process. Specific operations and/or tasks associated with the work are increasingly absorbed by the technology.

Given this relationship between conventional skills, abstract cognitive processes, the automation of tasks and the content of work, the issues of de-skilling and re-skilling need to be reconceptualized. The findings of the four occupations studied here conclude that the effects have been contra-dictory, often embodying both tendencies, and that they remain 'unfinished processes', continually shaped by the social relations of production.

The labour process is affected by both short and long term trends which may exert contradictory influences on skills and occupations. The immediate impact of computerization is, in its most visible and direct form, the automation of many of the manual tasks and skills which have historically defined the work of machinists, drafters and design drafters, in particular. Other short term trends include: the simplification of tasks, shifts in the demarcations between occupations around these tasks; and the acquisition of knowledge of computer-mediated symbols.

Long-term trends affecting the labour process, however, are located, not within the changes technology introduces in the production process, but rather in the social organization within which the technology is implemented. While computerization provides a range of possibilities to structure tasks and processes, it is the social organization which ultimately defines what those formations will be. As the data indicate here, these trends include the merger of job classifications within the shop, the redefinition of computerized technology from an automated to a conventional machine, the routinization of drafting work, and changes in the promotion ladder within the drafting and design occupations. These trends are related to automation in that the technology provides a range of choices regarding its use. However, they are primarily organizational decisions, not technical imperatives which re-shape and restructure the labour process.

Cockburn's[1] work is particularly useful here to articulate those aspects of skill which are affected by automation and the ways in which they are used in NC and Hurco work. Changes in the use of skill are compared using the three elements which comprize her definition of skill: (1) the skill that resides in the worker; (2) skill demanded by the job; and (3) the political definition of skill.

Comparing the changes in the four occupations studied reveals both some of the common effects of automation on skills and tasks, and the different ways in which the labour process is reorganized and skills utilized. Given the diverse nature of the skills, tasks, tools and technology involved in each of the occupations, these processes are affected differently. Therefore, the issues raised about the impact of automation on skills must carefully consider the following processes: the nature and range of physical and mental processes which comprise a task or set of tasks; the nature of conventional and automated tools, techniques used in production and design work; and the nature of skills applied to the automated labour process. The issue of skill is one aspect of changes facing workers in automated jobs.

Another issue which emerged in this study is the loss – or potential loss – of work, either directly because of computerization of the work, or by the performance of tasks which had been automated and simplified by related occupations. The ease with which computerized technology performs complex tasks permits work to be shifted more easily between occupations. As the data indicate, process engineers and design drafters face both a real and a potential loss of work as access to information by other occupations increases. This pattern – more than the possibilities of de-skilling – is emerging as a threat to these occupations.

In all of the occupations, automation replaced manual processes and performed some of the complex and time-consuming mathematical calculations required in many cases. The nature of this information-based technology automatically transfers manual and mathematical functions into the system. However, key elements of skill remain central to each occupation.

As we have seen, machinists are required to have extensive knowledge of machine tool capabilities and machining processes, regardless of its conventional or automated format. The nature of machine work and the division of labour in the shop at UFC which relies on skilled machinists make it impossible to eliminate or reduce the need for this expertise. Because of the complex variables that comprise the machining process, skills associated with this work cannot vary to any great extent. As far as the labour process itself, it appears that there will be little change in the direct application of skill and knowledge. Working on computerized machine tools will continue to demand highly qualified and experienced machinists.

Using the Office of Technology Assessment's categorization of components of skill (time to proficiency and judgment) to analyze the impact of CNC work on machinists' skills yields some interesting findings. Manual skills require years of practical experience (as opposed to formal education) for a machinist to be considered a skilled tradesperson. The automation of manual processes, in a sense, homogenizes skilled machining. Those manual skills and the artistry of craft work are made redundant, eliminating distinctions between those who have mastered the skills and machinists who are less proficient.

Judgment, however, is not affected by automation in quite the same manner for two reasons. First, conceptual skills are necessary to plan the process. Second, the machinist has to translate previously honed sensory skills into mathematical calculations. Not only does judgment remain a critical dimension of skill – it actually increases in importance and in its sophisticated application to the machining process.

Shifts in skill content embody contradictory tendencies. On the one hand, complex manual skills are eliminated along with the perceptual skills associated with performing these tasks. On the other hand, the mental work involved in programming demands more extensive planning and creative application of traditional skills, as well as the ability to translate the information sensory skills provide into programmable data.

Specific job hierarchies and occupational categories give form to the tasks and skills affected by automation. The trends exhibited by shifts in skill content take on particular organizational arrangements which are influenced by a number of factors: the existing division of labour and labour relations, the availability of trained personnel, the extent of automation, and the production needs of the firm. Job content is neither singularly nor primarily determined by automation. The reorganization of tasks and occupations is a managerial and, therefore, social decision.

Nevertheless, the merger of job classifications in the shop does pose a direct threat to the control machinists traditionally wield over the use of their skills. Change in work rules eliminates the categories of specialization in the trade and gives management an opportunity to exert greater control on the shopfloor. This reorganization also removes some of the power of the shop steward on the floor to negotiate the use of labour.

The displacement and creation of tasks outlined in the study conducted by the Office of Technology Assessment is evident in the shifting lines of demarcation between machining and methods at UFC. Tasks are absorbed into the technology through the development of software systems capable of transforming specialized computer programming skills into user-friendly data input. However, the transfer of tasks to the system also provides its increased access by machinists who use this information to broaden their role in the production process.

The automation of machine tools is certainly a catalyst for such reorganization to occur. Its ability to produce many different kinds of parts by virtue of the powerful programming capabilities of the equipment promotes a trend toward the homogenization of skills. However, management's decision to reclassify state of the art computer technology as a conventional machine tool challenges a purely technical explanation for changes in the organization of work. This reclassification is an organizational decision which allows any machinist – regardless of prior specialization and seniority – to operate it.

The outcome of these decisions is threefold: it cheapens the cost of labour by using less senior machinists on sophisticated computerized equipment; it accelerates the trend toward the homogenization of skills since all machine work can now be performed by any machinist; and it weakens the control machinists have traditionally exerted on the floor in the use of their skills.

The changes affecting the machinists at UFC are one strand of the modifications taking place in the occupation at large. It appears that UFC will continue to need highly skilled machinists although the efficiency of automated processes may ultimately mean that fewer machinists are required to do the same amount of work. However, this must be placed in the larger context of trends within the occupation.

Increasingly, there is a polarization between highly skilled machinists who can program and operate advanced computerized machine tools, and lesser skilled operators who monitor jobs in progress. The acceleration of

the trend toward polarization will eventually result in a two-tier occupation. If the skilled machinists' trade is to survive the automation
of its work, training will have to emphasize the skills and processes
associated with abstract problem-solving and the scientific and mathematical principles of machining. Efforts by labour unions to ensure a
well-designed training programme would need to articulate the necessary
skills to be maintained and those to be developed within the context of a
comprehensive curriculum.

The occupation of process engineer, as an historical outcome of
Taylorism, has a firm position in the automated workplace. NC programming work, in particular, is the cornerstone of the future of this
occupation, and the APT programming language serves as its distinctive
tool. Because of the lack of occupational specialization, and the complex
nature of production at UFC, process engineers are responsible for the entire
planning process, making it extremely difficult to fragment and routinize
the work.

Process engineers continue to direct their extensive knowledge and skills
to planning and programming on CAD/CAM. However, the simplification
of elements of the work and the capability of the system to store
information, creates the possibility that individual tasks can be redefined
toward conceptual categories of work. This could mean a rearrangement
of job classifications, particularly if the CAD/CAM system is a continuous
operation, linking each phase of design and production.

The arbitration ruling involving the use of the Hurco contributes to
the reorganization of methods' work. The elimination of the need for
specific programming skills and extensive knowledge of a programming
language, along with the availability of highly skilled machinists, created
an opportunity to redefine certain labour processes.

Process engineers do face some potential changes. In large, mass
production facilities, their work tends to be highly specialized. As generations of CAD/CAM become more sophisticated and widespread, the
need for such types of specialization will decline. The integrative nature
of the technology permits a wide range of applications and could lead
to a complete restructuring of operations. Speculation on future changes
includes the possibility of engineers programming and operating machine
tools from their computer terminals. Given the conditions under which
automation is introduced in most workplaces, this scenario is not likely
to occur in the near future. The continued reliance on NC equipment in
US industries, and the need for technical programming skills, indicates
that this occupation does not face a significant threat from this generation
of automated machine tools.

Out of the four occupations studied, drafters face an uncertain future. The productivity increases CAD is capable of generating, along with its integrative capacities, curtails both the growth of the occupation and the ability of drafters to gain necessary drafting and design skills thereby blocking upward mobility into design drafting.

The main work of this occupation is to create finished drawings. The ability of CAD to produce fairly detailed drawings bypasses the need for drafters in the numbers that currently exist. Moreover, the amount of detail and the information provided by the system makes it possible for other occupations to complete the work.

Using Cockburn's model, the 'match of skills' that occur as drafters are trained on CAD increases the potential for drafting work to become vulnerable to simplification, routinization and fragmentation, especially since drafters have not yet mastered their craft. Time to proficiency and the exercise of judgment is altered since CAD stores all of the relevant data needed to produce a design and detail it.

The potential for the total elimination – or at the least for a significant reduction – of drafters is unlikely at UFC for the near future for two main reasons: the complex nature of the designs that still require detailing experience, and the mechanism for promotion into design work is based on considerable drafting experience. Since UFC does not hire designers from outside of the company, some means of gaining this experience will have to continue on-site. Moreover, the union would resist any attempts to eliminate drafters and the current policy of internal promotions.

However, this may not be the case for other firms where detail work is less complex and more routinized. Therefore, changes in the occupation must be understood in relation to both a potential loss of work along with de-skilling by the automation of detail work and the processes associated with design apprenticeship. At this juncture, the long-term effects of CAD on drafting appear to be mostly negative. The measure of this impact will depend upon the drafters' ability to gain access to intensive CAD training for design work and, in union shops, the ability to hold the line against fragmentation and routinization.

As designers of engineering concepts and part of the drafting department, design drafters' skills encompass conceptual work and the practical application of ideas. The high degree of skills and level of experience of workers in this occupation provide a degree of security as design work is automated.

In those mass production industries where design work is fairly routinized, standardized and specialized, CAD may extend this process. The divisions between the conceptual or 'creative' aspect of engineering and its concrete image is blurred as the technology offers a nearly complete representation of a part or design. UFC's product lines, internal promotion policies and relatively general division of labour increase the likelihood that designers will maintain control over the use of skills in CAD work. However, productivity increases with automated design are significant and will have considerable impact on the long-term trends in the employment of design drafters. The capabilities of CAD to generate highly complex designs and perform many labour-intensive processes quickly will mean the need for fewer designers.

In cases where design work has already been fragmented or routinized (or CAD contributes to this process) work measurement techniques can, for the first time, be applied to conceptual work. UFC's attempt to do this has been met by resistance from designers who have traditionally controlled the design process and have transferred much of this control to CAD work. To the extent that management is able to institute quantitative measurements or break down the design process into a discrete set of tasks, the occupation will undergo dramatic changes.

The critical component of skill for designers, therefore, rests in its political dimension. It is in this arena that design work is defined and occupations delineated. The specific tasks associated with traditional job categories and descriptions are increasingly becoming inappropriate with information-based technology. In the past, the union has challenged encroachment into design work. The implementation of CAD into this process raises additional issues for the union surrounding the use of CAD as a source of information, a tool to complete specific tasks, and in relation to general design processes. Although the work of design drafters is influenced by automated technology, it is the redefinition of those tasks and processes which have been most affected that will determine the content of the occupation and the division of labour. The arena of labour relations and shopfloor politics will, ultimately, shape the skills used and the occupations involved in automated design.

Design equipment is relatively inexpensive. The ease with which CAD can be installed may mean that design work will be automated more quickly and systematically than production, making productivity increases in this area greater.

The conclusions presented here support the thesis that the labour process is subject to different tendencies, including de-skilling, re-skilling, the loss of work and changes in the use of skills. These tendencies are subject to negotiations and conflict over their meaning and expression in the labour process. Analysis of changes in the labour process need to be reconstituted to reflect these different tendencies and their influences on skills, going beyond the current debates which focus on either de-skilling or upgrading.

These findings contribute to restructuring this debate. The data indicate that while these processes may be occurring there is, simultaneously, a *transformation* of traditional manual skills, experience and knowledge into *conceptual* processes. This suggests that a definition of skill include abstract decision-making and conceptual transformations which now define skill, its use in production and in the reorganization of work.

Similarly, this increasingly conceptual process expands Cockburn's definition of skill to include the transformation of skills rather than their elimination through automation. The issues of skill and control now must also take into account the ways in which workers exert traditional knowledge and information within an automated context as well as those processes which are threatened by automation.

The failure of QCs at UFC can be explained on two levels: (1) workers' dissatisfaction with the process and outcome, and (2) management's partial and inconsistent support for the programme. The ideological issues surrounding participation is a key factor which accounts for workers' resistance and resentment toward management's decisions.

The dominant theme associated with Quality Circles and other worker participation schemes draws on the rhetoric of workplace democracy. Workers and their unions are presented with a plan that lists the mutual benefits of joint participation, packaged within a claim to extend democracy into the workplace. The promise of cooperative relations ignores the underlying conflict between labour and capital, namely, the subordination of labour within capitalist social relations.

Democracy refers to all parties having equal representation, access to all relevant information and the resources with which to institute change. As we have seen, the structure of QC includes the involvement of workers and union leadership at the lower end of the process with no actual power to affect decision-making. The QC apparatus has not replaced hierarchical relations nor does it provide workers with the necessary means to exercise authority to implement suggestions.

Appealing to workers on the grounds of democratic participation creates essentially a 'no-win' situation. Workers act in the belief that their participation includes the mechanism to determine outcomes. Two primary outcomes inherent in QC and other participation programmes are that workers will act in their own self-interest in addressing issues and suggesting solutions; and by defining the structure and terms of participation, management deliberately limits workers' role in decision-making and the impact decisions made in QC might have on managerial prerogatives.

These contradictory processes ensure conflicts over the meaning of participation and the nature and substance of issues raised within this context. The results can be seen in the high failure rates of most voluntary QC and other QWL programmes.

Beyond the ideology of democratic participation is a much different reason for attempting to engage workers in these programmes. The failure of Taylorism to encourage and maintain workers' cooperation, the capabilities associated with computerization that promote new forms of work organization, and an increasingly interdependent, global economy are forcing corporations to undergo dramatic changes in the way business is conducted and have set the stage for the restructuring of industrial labour relations. Organizations have responded to these external and internal changes by streamlining and decentralizing operations, reorganizing production methods and instituting new mechanisms for coordinating production and achieving labour's cooperation in this process.

Management's overriding concern is the flexible use of skills and the critical role cooperation assumes in creating and maintaining work organization forms that will sustain and reinforce that flexibility. The merger of job classifications in the shop and the attempts by management to encourage designers to quantify CAD work into a series of identifiable tasks are also part of the process designed to increase flexibility in the production process. Moreover, these organizational changes reduce workers' control over the use of their skills and gives management a greater opportunity to further reorganize the labour process.

Programmes like QC provide management with a framework to introduce and solicit ideas around productivity and flexibility within a context of cooperative relations and 'mutual benefit' to workers and management. With the appeal toward forming 'new' labour relations and acknowledging the importance of workers' expertise, management has developed a structure for incorporating suggestions without altering the prevailing hierarchical organization. Moreover, this structure strengthens the existing bureaucracy by acting as a conduit between central planning and increasingly decentralized operations for a constant exchange of information and to instruct lower level management on changes which need to be implemented.

The organization of work is part of the overall corporate restructuring currently underway. Less associated with specific tasks and existing occupational hierarchies and classifications, the emerging forms of work and decision-making are linked with informational processes, increased access to data and more generalized applications of knowledge than currently exist.

The transformations taking place have been swift and dramatic in some industries, gradual and incremental in most. Industries which have instituted these changes in more dramatic forms are those that have been most acutely affected by economic developments of the past two decades. These responses are the origins of a new system of production.

While management continues to re-shape the work process around computerized production methods, unions have developed some strategies of their own to deal with its effects on jobs and collective bargaining. The International Metalworkers Federation (IMF) appealed to unions around the world to 'notify (them) of any "extraordinary activity" in their plants' indicating that work is being relocated from striking plants.[2] The technology which allows work to be electronically transferred from one plant to another also increases the potential for a movement toward international labour solidarity. Domestically, unions have used the concept of QWL to fight plant shutdowns and increase participation in production. The United Electrical Workers enlisted the aid of the community and local government in fighting the closure of Morse Cutting Tools. The prospect of job loss and social dislocation within the community prompted the city to threaten Gulf and Western (the parent company) with a declaration of the right of 'eminent domain' if Morse was to be closed. In another case, to attract new buyers, workers at a GE steam-turbine plant scheduled to close developed a list of environmental products which could be successfully produced.[3]

The use of computerized automation and the institution of QWL in the labour process are part of the historical continuation to control production, enlist labour cooperation, and reduce uncertainty within the marketplace. These changes are also part of the current shift in capitalist development and reflect both the historical continuation of struggle between labour and capital, and the specificity of this struggle as it relates to the emergence of a new global economic order.

The computerization of the labour process and the implementation of neo-human relations programmes such as Quality Circles go beyond Taylorist strategies in two significant ways. First, the reintegration of conception and execution, the elimination of *specific* sets of tasks in favour of informational processes and, in some cases, the redesign of the labour process into 'work teams' removes traditional barriers between occupations. Second, along with this reintegration, worker participation programmes attempt to engage workers directly in their own supervision in terms of productivity and efficiency. The notion of 'non-adversarial' relations and cooperation around common goals are attempts at reducing resistance and integrating workers into an increasingly interdependent production process. This is accomplished, in part, by soliciting and implementing their recommendations for improving productivity. The movement from direct supervision and toward self-supervision is part of the mechanism of increased integration into the production process. Although computerization and worker participation programmes are not always implemented together, both emphasize reintegration of work processes and self-directed work relations. In these respects, computerization and worker participation schemes are a break with Taylorism.

Labour unions faced a crisis at the turn of the century as capital consolidated into corporations, introducing mass production techniques, scientific management and new forms of supervision which threatened the existence of skilled crafts and their control over the labour process. Contemporary unions are now confronted with a similar dilemma. However, the global and technological conditions which allow corporations to electronically transfer work to virtually anywhere in the world create their own contradictions.

Workers can significantly sabotage attempts at the global transfer of work by fostering links among unions worldwide. Union solidarity networks on the national and international levels combined with new strategies in collective bargaining that will challenge management's campaign for major concessions and the elimination of job security provisions are approaches currently being explored within the labour movement. These actions, as much as computerization and corporate restructuring, will determine the future of the workplace.

Appendix 1

INTERVIEW SCHEDULE

Before we begin, I want to remind you, once again, that this interview is confidential. Your name and identity will not be revealed. Do you have any questions about the interview or the study before we start?

What is your occupation? Age? Seniority? The department or division you work in? How many years have you worked at (*respondent's occupation*)? What type of training did you receive? Have you worked at any other occupation at —————? (If so, what was it? How long were you a —————?

What type of automated equipment do you work with? How long has your work been automated? What training was involved to learn how to operate —————? How much of it was on the job training? How much of it was spent outside of the company in classes or course work?

Who trained you? ————— personnel? the manufacturer of the equipment?

Now, I'd like to talk about your job as a (machinist, process engineer, drafter, design drafter)?

What is involved in doing conventional (machining, planning, drafting, designing)?

Could you describe what is involved in —————?

Probe: Talk about (specific process). Probe: What specific skills do you need to do —————? Probe: Why is this important for performing —————? Do you ever need to discuss or consult with another co-worker or someone in another occupation about —————? Are there any particular aspects of conventional work you enjoy most? What are they?

What about the work is most satisfying?

What has changed about your work since it has been automated? Probe: Could you describe what is involved in —————? Probe: How is this different from —————?

Are there particular things you like about working with (NC, Hurco, CAD/CAM, APT,)?

Is there anything you dislike about working with (NC, Hurco, CAD/CAM, APT)?

Probe: Could you describe what is different about doing —————? Are there things you can do now that were not possible prior to automation?

On the other hand, are there things that you once were able to do that cannot be done because of automation?

212

Do you feel that the use of this equipment has increased or decreased the use of your skills? Probe: Describe how it has changed ——————?

Have you found that automation allows you to combine several operations rather than performing separate tasks? Could you describe them and how they have changed?

Additional probe: Earlier you mentioned (specific process). What you're mentioning now is a bit different. Could you explain how you see these differences (or changes)?

Have you found that because of automation, parts of your work can be performed by someone in another occupation? (If yes) describe the work and the occupations involved?

Have you found that automation could allow you to perform other job functions not normally covered by your occupation? Could you describe the work and the occupations involved?

We are finished with the part of the interview that concerns automation. Are there any other issues you see as important that we have not covered?

Next, I would like to talk about the Quality Circle program.

QC Participants
Why did you join?
How long have you been a member?
How long has the Circle been operating?
Could you describe the training program you took part in to qualify for participating in QC.
Probe: Could you discuss (particular experience or event)?
Probe: What did you learn from (particular experience)?
What did you like/dislike about the training program?
How does the circle work? What is a typical meeting like?
Probe: Could you talk about (*process*)? Probe: How do you decide ——————?
Who runs the meetings?
What kinds of decisions does —————— make?
Who approves the proposals made in the circle?
What proposals have been accepted? rejected?
Who takes the minutes at the meetings?
Probe: Could you discuss the issues surrounding the disagreements over the minutes?
What is the criteria use in accepting/rejected a proposal?
Are there instances where you disagree with another worker's proposal? What was the issue? How did other circle members react?
Does management ever present proposals to the circle?
(If so) What are they? What do you think about (*proposal*)? How is dissent or disagreement handled within the circle?
Probe: Could you describe what happened? (events leading up to the disagreement and workers' responses to it)

What reactions do you get from workers on the floor who are not in the program?

Have you ever suggested an idea or proposal for the circle to consider? (If so) What was the outcome?

Are proposals generally made by individuals or do workers, as a group, decide on a particular idea or an agenda?

Probe: Earlier you mentioned (*specific process/conflict in the circle or incident*). What you just mentioned is something a bit different. Could you explain how you see these differences?

Do you get together (as union members) outside of the circle to discuss the issues that were raised? (If so) Could you talk about what is discussed?

Prior to joining the circle, were you active in your local? Have you become more/less active as a result of participating in the program?

Why do you think it is important for workers to participate in quality circles?

How do you see the role of the union in the program?

Does the leadership actively encourage participation?

Are there any issues the union would like to put on the agenda (or has put on the agenda)?

Participants who dropped out of the program
How long were you a member?
Why did you join?
Why did you decide to leave?
Probe: Could you discuss (*the incident*)? (At this point, ask other relevant questions from the above section which would help clarify the responses).
Do you think the union should continue to support quality circles? Why?

Workers who never joined QC
Why didn't you join?
What do you think of the suggestions that come out of the circle?
What changes have you seen since the program started?
Which particular ones do you support? reject?
At this point, are there any circumstances under which you might consider joining a circle?

All Respondents
Have you noticed changes in labor-management relations because of workers' involvement in the quality circle program?
Probe: Could you discuss *event*?

This concludes the interview. Is there anything you would like to add that wasn't discussed in the interview that you feel is important? Do you have any other questions you would like to ask me?
Thank you for taking the time to participate in this project.

Appendix II

QUALITY CIRCLE CHART

Categories of QC issues	Issues raised in QC	Encouraged by management	Approved but not implemented (or implemented differently from workers' experctations)	Rejected or resisted by management	Implemented as suggested	Union intervention
	photoplotter		X			
	typesetter odour			X		
Health and safety	chemical			X		X
	safety switches			X		
	room arrangements	X				
	noise			X		
	spray booth	*unsure of status at time of interview*				

Category	Item				
Cost-savings and productivity	keypunch elimination				X
	OS on file				X
	microfilm tool design				X
	OS information on dimensioning				
	tool books		X		
	coputerized production of file system			X	
	new tools and uniform rollaways		X		
Managerial issues	eliminate board work	X			
	create master design book	X			
	taping CAD jobs	X			
	drafters performing design work	X			
	tape storage/ password on computer/CAD room arrangement	X			

union attempted to change procedures for promotion into design

Notes

1 Introduction

1. Quality of Work Life (QWL) is an 'umbrella' term covering a wide range of management-instituted programmes. Although the types of programmes differ in structure and style, they are primarily concerned with increasing productivity, product quality and cooperation of the workforce in production. The distinctions between these programmes are important in that each is designed to address particular programmes within different types of production and service sector industries. However, this book will explore one type, quality circles.
2. Daniel Nelson, *Managers and Workers: Origins of the New Factory System in the U.S. 1880–1920* (Madison, Wisconsin: University of Wisconsin Press, 1975), p. 25.
3. The initials stand for 'Unidentified Flying Corporation' and should not to be confused with any corporation which may have the same initials.
4. Harry Braverman, *Labor and Monopoly Capital* (New York: Monthly Review Press, 1974).
5. Robert Blauner, *Alienation and Freedom* (Chicago: University of Chicago Press, 1964).
6. Paul Thompson, *The Nature of Work: An Introduction to Debates on the Labour Process* (London: Macmillan Press, 1983).
7. P. Goldman and D. Van Houten, 'Managerial Strategies and the Worker', in K. Benson (ed.), *Organizational Analysis: Critique and Innovation*, (New York: Sage, 1977), pp. 112.
8. John Kelly, 'Management's Redesign of Work: Labour Process, Labour Markets and Product Markets', in David Knights, Hugh Willmott and David Collinson (eds), *Job Redesign* (Brookfield, Vermont: Gower Publishing, 1985), pp. 30–51.
9. Richard Edwards, *Contested Terrain*.
10. Paul Thompson, *The Nature of Work*, p. 152.
11. Daniel Nelson, *Managers and Workers*.
12. Carol Haddad, 'Technological Change and Reindustrialization: Implications for Organized Labor', in Donald Kennedy (ed.), *Labor and Reindustrialization: Workers and Corporate Change* (Department of Labor Studies: Pennsylvania State University, 1984), pp. 137–66.
13. Donald H. Sanders, *Computers and Management* (New York: McGraw-Hill, 1970), pp. 124–25.
14. Herbert Simon, *The New Science of Management Decision*, revised edition (New Jersey: Prentice Hall, 1977); Peter M. Blau, 'Decentralization in Bureaucracies', in Mayer N. Zald (ed.), *Power in Organizations*, (Nashville, Tennessee: Vanderbilt University Press, 1970), pp. 150–74.

217

15. Raymond Dreyfack, *Making it in Management – The Japanese Way* (Rockville Center, N.Y.: Farnsworth, 1982); John Simmons and William Mares, *Working Together* (New York: Knopf, 1983); and William Ouchi, *Theory Z: How American Business Can Meet the Japanese Challenge* (Reading, Mass.: Addison-Wesley, 1981).

16. *Business Week*, 11 May 1981; Delmar L. Landen, 'The Real Issue: Human Dignity', *The Work Life Review* (November, 1982); Richard E. Walton, 'How to Counter Alienation in the Plant', *Harvard Business Review* (November/December, 1972): 70–81.

17. Edward Vaughan, 'Industrial Democracy: Consensus and Confusion', *Industrial Relations Journal*, Vol. 11, No. 1 (March/April, 1980): 50–6.

18. John F. Witte, *Democracy, Authority and Alienation in Work: Workers' Participation in an American Corporation* (Chicago: University of Chicago Press, 1980).

19. U.S. Department of Labor, Labor-Management Services Administration, 'Starting Labor-Management Quality of Work Life Programs' (Superintendent of Documents, U.S. Government Printing Office: Washington, D.C., September, 1982).

20. Ibid.

21. Irving Siegel and Edgar Weinberg, *Labor-Management Cooperation: The American Experience* (Kalamazoo, Michigan: W. E. Upjohn Institute for Employment Research, 1982).

22. Charles Hecksher, 'Worker Participation and Management Control', *Journal of Social Reconstruction* (January/March, 1980).

23. Robert Wrenn, 'Management and Work Humanization', *The Insurgent Sociologist*, Vol. XI, No. 3 (Fall, 1982): 23–38.

24. Daniel Yankelovich and John Immerwahr, 'Putting the Work Ethic to Work', *Society* (January/February, 1984): 58–76.

25. Mick Marchington, *Responses to Participation at Work* (Westmead, England: Gower Publishing, 1980); Harvie Ramsey, 'Cycles of Control: Worker Participation in Sociological and Historical Perspective', *Sociology*, Vol. 11, No. 3 (September, 1977): 481–506.

26. Harley Shaiken, *Work Transformed*, p. 18.

27. For example, General Motors has used data on productivity output in its parts plants to raise work quotas. Called 'whipsawing' by unions, plants were expected to compete with one another to increase productivity with the threat of plant closures or outsourcing.

28. Rosabeth Moss Kanter, *The Change Masters* (New York: Simon and Schuster, 1983), p. 411.

29. Ibid., pp. 52–3.

30. *The New York Times*, 21 April 1985.

31. The foreman's role, in particular, has increased in importance as corporations decentralize authority and decision-making.

32. This applies to relationships within levels of management as well, particularly in view of the fact that horizontal decision-making increases within a parallel decision-making structure.

33. Peter Drucker, *Technology, Management and Society* (New York: Harper and Row, 1970), p. 47.
34. The term methods 'man' is used in the shop. The text will refer to both titles, although the gender reference will not be used.

2 Site Description and Methodology

1. Paul Goldan and Donald Van Houten, 'Managerial Strategies and the Worker', in Kenneth benson (ed.), *Organizational Analysis: Critique and Innovation* (Beverley Hills: Sage, 1977).
2. Wolf V. Heydebrand, 'Organizational Contradictions in Public Bureaucracies: Toward a Marxian Theory of Organizations', ibid., p. 87.
3. The populations from the samples were drawn are as follows: machinists – 108; process engineers – 30; drafters – 40; design drafters – 30; and sheet metal workers – 35.

3 Review of the Literature

1. Harry Braverman, *Labor and Monopoly Capital*, p. 58.
2. These variables were: '(a) the quality of the metal which is to be cut; (b) the diameter of the work; (c) the depth of the cut; (d) the thickness of the shaving; (e) the elasticity of the work and of the tool; (f) the shape or contour of the cutting edge of the tool, together with its clearance and lip angles; (g) the chemical composition of the steel from which the tool is made, and the heat treatment of the tool; (h) whether a copious stream of water or other cooling medium is used on the tool; (i) the duration of the cut, i.e., the time which a tool must last under pressure of the shaving without being reground; (j) the pressure of the chip or shaving upon the tool; (k) the changes of speed and feed possible in the lathe; (l) the pulling and feeding power of the lathe.' Quoted from F. W. Taylor, *The Art of Cutting Metals*, p. 32 in Frank B. Copley, *F. W. Taylor: Father of Scientific Management*, Vol. 1 (New York: Augusten Kelley Publishing, 1969), p. 245.
3. Charles Walker and Robert Guest, *Man on the Assembly Line* (Cambridge, Mass.: Harvard University Press, 1952); Eli Chinoy, *Automobile Workers and the American Dream* (New York: Doubleday, 1955).
4. Robert Blauner, *Alienation and Freedom*.
5. Michael Burawoy, 'Toward a Marxist Theory of the Labor Process', *Politics and Society*, Vol. 8 (1978): 288.
6. Ibid.
7. Tony Manwaring and Stephen Wood, 'The Ghost in the Machine: Tacit Skills in the Labor Process', *Socialist Review* 74 (March–April 1984): 56.
8. Karl Marx, *Capital*, Vol. 1 (New York: International Publishers, 1975).
9. Paul Thompson, *The Nature of Work*, p. 57.

10. F. Child, 'Managerial Strategies, New Technology and the Labour Process', *Job Redesign*, D. Knights, H. Willmott, and D. Collinson (eds), p. 112.

11. David F. Noble, *Forces of Production: A Social History of Industrial Automation* (New York: Oxford University Press, 1986).

12. Harley Shaiken, *Work Transformed*.

13. Paul Adler, 'Technology and Us', *Socialist Review* 85 (January–February 1986): 67–96.

14. Larry Hirschhorn, *Beyond Mechanization* (Cambridge, Mass.: MIT Press, 1983).

15. Noble provides an excellent historical account of the state supported and subsidized development of NC and APT and its use in defence and other state-subsidized industries – aircraft, aircraft engines and parts and the machine tool industry. Businesses which did not have the cushion of military contracts or state support found it difficult to automate their shops. The cost of NC equipment, inflated by its military support, plus its complex operations and unreliability as a technically advanced and accurate system, limited demand outside of defence industries. According to Noble, these conditions contributed to labour costs outpacing capital expenditures for new equipment by the mid-1960s. The 'revolution' in computerized machining predicted by industry analysts in the early post-War years did not extend far beyond DOD contracting firms.

16. *American Machinist*, 25 October 1954; and *Business Week*, 14 March 1959.

17. *Business Week*, 14 March 1959.

18. Experience with problems facing corporations using NC led a few to comment on the short-sightedness of this approach. Noble quotes presidents of machine tool manufacturing firms from the 1950s and the 1980s on the critical importance of skilled labour in NC machining. See Noble, *Forces of Production*, pp. 236–7 and p. 338 for their comments.

19. David Noble, *Forces of Production*, p. 221.

20. Frank B. Copely, *F. W. Taylor: Father of Scientific Management*, Vol. 1.

21. Larry Hirschhorn, *Beyond Mechanization*, p. 4.

22. Paul Adler, 'Technology and Us', *Socialist Review*, p. 80.

23. Robert Blauner, *Alienation and Freedom*.

24. Cynthia Cockburn, *Brothers: Male Dominance and Technological Change* (London: Pluto Press, 1983).

25. Ibid., p. 113.

26. Ibid.

27. Ibid.

28. *Computerized Manufacturing Automation: Employment, Education, and the Workplace* (Washington, D.C.: US Congress, Office of Technology Assessment, OTA-CIT-235, April 1984).

29. Ibid., p. 110.

30. Ibid., pp. 110–11.

4   Automation on the Shopfloor: Machinists

1. Harley, Shaiken, *Work Transformed*
2. See Shaiken for a much richer description of machine shopfloor settings in the United States, Chapter 2.
3. Harley Shaiken, *Work Transformed*, p. 66.
4. Office of Technology Assessment, *Computerized Manufacturing Automation*, p. 136–7.
5. Ibid.
6. Roger Tulin, *A Machinist's Semi-Automated Life* (San Pedro: Singlejack Books, 1984), pp. 14–15.
7. This issue of jurisdictional dispute between machinists and methods will be analyzed in greater detail later in Chapter 5.
8. Harley Shaiken, *Work Transformed*, p. 78.
9. Harley Shaiken, *Work Transformed*; and David Noble, *Forces of Production*.
10. As we will see later in this chapter, the Hurco opens up the possibility to reverse this process.
11. Paul Thompson and Eddie Bannon, *Working the System: The Shopfloor and New Technology* (London: Pluto Press, 1985) distinguish between technical and subjective de-skilling. Technical de-skilling refers to the actual skills which are affected by mechanization. Subjective de-skilling addresses how these changes have actually been experienced by workers. According to Thompson and Bannon, those experiences will influence the patterns of conflict and resistance over technical de-skilling. They point out that there are wide variations in each category.
12. Shoshana Zuboff, 'New Worlds of Computer-Mediated Work', *Harvard Business Review* (September–October 1982): 142–52.
13. This issue will be explored in detail in Chapter 5 when the arbitration ruling on the Hurco will be analyzed.
14. Office of Technology Assessment, *Computerized Manufacturing Automation: Employment, Education, and the Workplace*, p. 111.
15. Paul Thompson, *The Nature of Work*.

5   Merger of Job Classifications

1. Although this work does not detail the work and skills of the sheet metal department workers, some of the shop stewards and members were interviewed. Both sheet metal and machining areas are part of the same merger agreement. Their work was subject to the same NC and CNC automation with similar results – the elimination of specialized skills; the reliance on conceptual rather than manual skills; and the removal of shopfloor control from stewards and workers.

2. Shoshana Zuboff, 'New Worlds of Computer-Mediated Work', p. 146.
3. Robert Wrenn, 'Management and Work Humanization', *Insurgent Sociologist*, Vol. XI, No. 3 (Fall 1982): 23–38; Jeremy Brecher and David Montgomery, 'Crisis Economy: Born Again Labor Movement?', *Monthly Review* (March 1982): 1–17.
4. Internal union document, Local 8, IUE, 'Post-Hearing Memorandum to Arbitrator', Grievance M-1-81, p. 9.
5. Internal company document, 'LETTER OF INTENT regarding Use of the Manual Data Input (M.D.I.) i.e., Hurco Machine between the Company and Local 8.
6. Internal union document, 'Post-Hearing Memorandum', Grievance M-1-81, p. 18.
7. Internal Union Document, Grievance, 3rd Meeting, Local 8, IUE, 28 April 1981, p. 2. The union gave permission to the company to allow members of another local to alter, create or redesign harness designs involving not more than five wires.
8. Internal Union Document, 'Arbitration Ruling', 30 August 1982, p. 40.
9. Ibid, p. 40.

6   Process Engineers

1. They also prepare instruction sheets for sheet metal and assemblers. However, since this research focuses primarily on the machinists, the study of methods' work will be limited to its application to machining.
2. Frederick W. Taylor, *Scientific Management* (Westport, Ct.: Greenwood Press, 1972), p. 102.
3. Ibid., p. 166, figure 5.
4. Taylor did introduce different tools for calculations and machines were beginning to be motor rather than belt driven. However, these changes did not significantly alter the machining process itself.
5. Keep in mind that UFC also offers courses in APT, providing an opportunity for machinists to gain some fundamental understanding of NC programming.
6. The younger, less senior workers all have associates degrees from local technical colleges which offer a degree in manufacturing technologies which includes courses as: advanced mathematics physics, drafting, design and mechanics courses related to basic manufacturing applications.
7. Harley Shaiken, *Work Transformed*, p. 100.
8. Arthur D. Roberts and Richard C. Prentice, *Programming for Numerical Control Machines* (2nd edn; New York: McGraw-Hill, 1978).
9. Ibid., p. 9.
10. Shoshana Zuboff, 'New Worlds of Computer-Mediated Work', *Harvard Business Review*, p. 146.
11. Harry Braverman, *Labor and Monopoly Capital*; Andrew Zimbalist (ed.), *Case Studies on the Labor Process*; and Mike Cooley, *Architect or Bee?:*

The Human Technology Relationship (Boston: South End Press, 1980)
12. Certain factors need to be taken into account when making such projections, however. The influence of government contract support, for example, would affect the hiring and layoff policies of the corporation.
13. Chapter 5 addresses the issue of machinist programming in a much broader context of management's re-definition of CNC equipment as a conventional machine tool.
14. See Chapter 5 for a detailed analysis.

7 Automation and the Design Process: Drafters

1. UFC uses the Computervision CAD system and is referred to by its users as CADDS – computer-aided drafting and design system. Whenever a respondent refers to the system in the interviews, I will use Computervision's initials. My own references to computer-aided design work will use the generic term – CAD.
2. As noted earlier, UFC has not used the system to link the design and manufacturing processes. However, the capability does exist for such a connection.
3. The test has become a controversial issue within the union. Currently, to pass the test successfully it is mathematical knowledge, not the ability to do actual design work as in the past, which is the critical factor. Drafters who are the more recent graduates of technical schools have a better chance of passing this kind of test than a more senior drafter who has gained design experience at the board. This may indicate the declining importance of board skills in favor of CAD work. However, as design drafters indicate, knowledge of design remains indispensable on CAD.
4. Office of Technology Assessment, *Computerized Manufacturing Automation: Employment, Education, and the Workplace*, p. 43–4.
5. Shoshana Zuboff, 'New Worlds of Computer-Mediated Work', *Harvard Business Review*, p. 145.
6. Ibid, p. 146.
7. Harley Shaiken, *Work Transformed*; and Mike Cooley, *Architect or Bee?*
8. Some design drafters, on the other hand, are given release time to attend classes at a local community college. This will be discussed in Chapter 8.
9. The International Association of Machinists (IAM), faced with the introduction of massive technological change, drew up a document which establishes a framework for negotiating technology agreements. Called 'The Technology Bill of Rights' it outlines conditions under which technology could be implemented. One of these conditions included the provision that automation should enhance skill and expand workers' responsibilities. Such an agreement, although broad and general in scope, sets up parameters for the implementation of technology into the labour process.

10. This is not to state that a decline in the number of drafters is a purely technical outcome. At UFC, the continued and/or increase in support through government contracts and the nature of its products will also significantly influence the number of employed drafters. Several studies have noted that actual increases in the numbers of CAD drafters have occurred as product markets have expanded and new products introduced. Office of Technology Assessment, *Computerized Manufacturing Automation: Employment, Education and the Workplace*; and Harold Salzman, 'Computer Technology and the Automation of Skill', Paper presented at the Annual Meetings of the Eastern Sociological Society, New York, New York, March 1986.

11. Government Accounting Office, *Advances in Automation Prompt Concern Over Increased US Unemployment*, 25 May 1982, p. 33.

8 Design Drafters

1. Bureau of Employment Security, U.S. Department of Labor, *Dictionary of Occupational Titles* (3rd edn; 1965, Vol. 1), p. 218.
2. Harold Salzman, 'Computer Technology and the Automation of Skill', p. 44.
3. Ibid, p. 44.
4. Ibid, p. 44.
5. Barry Wilkinson, *Shopfloor Politics of New Technology* (London: Heinemann Books, 1983), pp. 31–2.
6. Mike Cooley, *Architect or Bee?*

9 Historical Overview and Review of the Literature

1. Giussepe de Lampedusa, *The Leopard*, translated from the Italian by Archibald Colquhoun (New York: Pantheon, 1960), p. 40
2. The National Industrial Conference Board, *Practical Experience with Profit-Sharing in Industrial Establishments*, Report 29 (Boston, Mass., June 1920), p. 6.
3. Alfred Chandler, Jr., *The Visible Hand: The Managerial Revolution in American Business* (Cambridge, Mass.: Belknap Press, 1977).
4. John Howell Harris, *The Right to Manage: Industrial Relations Policies of American Business in the 1940s* (Madison, Wisconsin: University of Wisconsin Press, 1982), p. 17.
5. Ibid, p. 17.
6. John H. Harris, *The Right to Manage*, p. 164.
7. Thomas A. Kochan, Harry C. Katz and Robert B. McKersie, *The Transformation of American Industrial Relations* (New York: Basic Books, 1986), p. 148.
8. Steve Fraser, 'Industrial Democracy in the 1980s', *Socialist Review*, Vol. 13, No. 6 (November/December, 1983): 11.

9. Report of a Special Task Force to the Secretary of Health, Education and Welfare, *Work in America* (Cambridge, Mass: MIT Press, 1973).

10. John Simmons and William Mares, *Working Together* (New York: Knopf, 1983); William Ouchi, *Theory Z: How American Business Can Meet the Japanese Challenge* (Reading, Mass.: Addison-Wesley, 1981); R. Katzell *et al.*, *A Guide to Worker Productivity Experiments in the U.S. 1971–75* (New York: NYU Press, 1977); and Douglas McGregor, *The Human Side of Enterprise* (New York: McGraw-Hill, 1960).

11. Walter R. Nord, 'The Failure of Current Applied Behavioral Science – A Marxian Perspective', *Journal of Applied Behavioral Science*, 10: 4 (October–December 1974), p. 576.

12. Ivar Berg, Marcia Freedman and Michael Freeman, *Managers and Work Reform: A Limited Engagement* (New York: The Free Press, 1978), p. 11. Emphasis in original.

13. John Simmons and William Mares, *Working Together*.

14. Robert J. Thomas, 'Participation and Control: New Trends in Labor Relations in the Auto Industry', Working Paper 315 Center for Research on Social Organization (Ann Arbor: University of Michigan), p. 13.

15. Michael Burawoy, *Manufacturing Consent: Changes in the Labor Process Under Monopoly Capitalism* (Chicago: University of Chicago Press, 1979).

16. Wolf V. Heydebrand, 'Technocratic Corporatism: Toward a Theory of Occupational and Organizational Transformation', New York University, unpublished paper.

17. Barry Stein and Rosabeth Moss Kanter, 'Building the Parallel Organization and Creating Mechanisms for Permanent Quality of Work Life', *Journal of Applied Behavioral Science*, Vol. 6 (1980) No. 3: 371–88.

18. Ibid, p. 386.

19. Thomas Kochan *et al.*, *The Transformation of American Industrial Relations*, p. 148.

20. Donald M. Wells, *Empty Promises: Quality of Work Life Programs and the Labor Movement* (New York: Monthly Review Press, 1987).

21. Thomas Kochan *et al.*, *The Transformation of American Industrial Relations*, p. 118.

22. Ibid, p. 120.

10 Quality Circles

1. Daniel Yankelovich and John Immerwahr, 'Putting the Work Ethic to Work', *Society*, p. 66.

2. Ibid, p. 66.

3. *The New York Times*, 'Week in Review', 11 September 1983, p. 3.

4. By 1983, UFC had approximately 400 Quality Circles operating throughout the corporation.

5. Internal company publication, *Quality Circle Report*, Vol. 1, No. 1 (March 1983): 3.

6. It is beyond the scope of this study to measure and assess the accuracy and strength of this relationship. Beginning with Elton Mayo and the Hawthorne study in the 1920s, industrial psychologists, sociologists and management scientists have attempted to identify those characteristics which improve or impede cooperation and productivity. Their results have often proved to be very inconclusive and, often, contradictory. Its relevance to this work lies mainly in the *ideological* justification for promoting these programmes and explaining its role within the organization.

7. John Simmons and William Mares, *Working Together*; William Ouchi, *Theory Z*.

8. John Dickson, 'Participation as a Means of Organizational Control', *Journal of Management Studies*, Vol. 18 (1981) No. 2: 159–76; Harvie Ramsey, 'Phantom Participation: Patterns of Power and Conflict', *Industrial Relations Journal*, Vol. 11, No. 3 (July/August 1980): 46–59.

9. Steve Fraser, 'Industrial Democracy in the 1980s', *Socialist Review*, 72 (1983), pp. 99–122.

10. Internal company document, UFC newsletter, February 1984, p. 4.

11. Internal company document, *Quality Circle Report*, UFC newsletter, October 1984, p. 7.

12. Ibid, p. 6.

13. Internal company document, UFC newsletter, p. 4.

14. Barry Stein and Rosabeth M. Kanter, 'Building the Parallel Organization and Creating Mechanisms for Permanent Quality of Working Life', *Journal of Applied Behavioral Science*, Vol. 16 (1980) No. 3: 371–2.

15. Ibid, p. 372.

16. Ibid, p. 372.

17. Daniel Yankelovich and John Immerwahr, 'Putting the Work Ethic to Work', *Society*; and John Simmons and William Mares, *Working Together*.

18. Barry Stein and Rosabeth Kanter, 'Building the Parallel Organization and Creating Mechanisms for Permanent Quality of Work Life', *Journal of Applied Behavioral Science*, p. 373.

19. *Quality Circle Members Manual*, Internal UFC document, p. 6.

20. Quoted in Jeremy Main, 'The Trouble with Managing Japanese Style', *Fortune*, 21 April 1984, p. 50.

21. QC Members Manual, UFC document, p. 146.

22. Mike Parker, *Inside the Circle: A Union Guide to QWL* (Boston: South End Press, 1985), p. 17.

23. Paul Goldman and Donald Van Houten, 'Uncertainty, Conflict and Labor Relations in the Modern Firm 1: Productivity and Capitalism's "Human Face"', *Economic and Industrial Democracy* Vol. 1, No. 1 (February 1980): 74.

24. John F. Witte, *Democracy, Authority and Alienation in Work: Workers' Participation in an American Corporation* (Chicago: University of Chicago Press, 1980), p. 2.

25. Robert J. Thomas, 'Participation and Control: New Trends in Labor Relations in the Auto Industry', Center for Research on Social Organization, University of Michigan, Ann Arbor, Working Paper 315.
26. William Glaberson, 'Is OSHA Falling Down on the Job?', *New York Times*, 2 August 1987, Section 3, p. 1.
27. Ken C. Kusterer, *Know-How on the Job: The Important Working Knowledge of 'Unskilled' Workers* (Boulder, Colorado: Westview Press 1978).
28. See Appendix 2.
29. Jeremy Main, 'The Trouble with Managing Japanese Style', *Fortune*; Mike Parker, *Inside the Circle*.
30. Harvie Ramsey, 'Phantom Participation: Patterns of Power and Conflict', *Industrial Relations Journal*, Vol. 11 (1980) No. 3: 46–59.
31. Paul Cathey, 'Industry's Man in the Middle', *Iron Age*, 2 January 1983, pp. 36–9.
32. Donald Wells, *Empty Promises*.
33. Since no interviews were conducted with QC management, it is not possible to determine the extent to which pressure was exerted to 'produce successes' in the programme. Nevertheless, as administrators of this highly visible experiment, no doubt they were subject to some measures of productivity and accountability by corporate executives overseeing the project.
34. Donald Wells, *Empty Promises*, p. 111.
35. Ibid, p. 69.
36. Ibid.
37. Ibid.

11 Conclusions

1 Cynthia Cockburn, *Brothers*.
2 Bureau of National Affairs, 'Metalworkers Federation Acts to Help Striking West German Auto Workers', June 4 1984, p. 691.
3 *The New York Times*, 'Unions Are Shifting Gears But Not Goals', 31 March 1985, p. 10.

Bibliography

Adler, Paul, 'Technology and Us', *Socialist Review* 85 (January / February 1986): 67–96.

American Machinist, 25 October 1954.

Berg, Ivar, Marcia Freedman and Michael Freeman, *Managers and Work Reform: A Limited Engagement* (New York: The Free Press, 1978)

Bernstein, Paul, 'Necessary Elements for Effective Worker Participation in Decision-Making', *Journal of Economic Issues* 10 (2 June 1976): 490–522.

Blau, Peter, 'Decentralization in Bureaucracies', in Mayer N. Zald (ed.), *Power in Organizations*, pp. 150–74 (Nashville, Tenn.: Vanderbilt University Press, 1970).

Blauner, Robert, *Alienation and Freedom* (Chicago: Chicago University Press, 1964).

Braverman, Harry, *Labor and Monopoly Capital* (New York: Monthly Review Press, 1974)

Brecher, Jeremy, and David Montgomery, 'Crisis Economy: Born Again Labor Movement?' *Monthly Review*, March 1982: 1–17.

Burawoy, Michael, *Manufacturing Consent: Changes in the Labor Process Under Monopoly Capitalism* (Chicago: University of Chicago Press, 1979).

—— 'Toward a Marxist Theory of the Labor Process', *Politics and Society* 8 (1978): 247–312.

Bureau of National Affairs, *Employee Relations Weekly*, 'Unions Face New Hurdles, Threats after "Milwaukee Spring Ruling"', 6 February 1984.

Cathey, Paul, 'Industry's Man in the Middle', *Iron Age* (2 January 1983): 36–9.

Chandler, Alfred, Jr., *The Visible Hand: The Managerial Revolution in American Business* (Cambridge, Mass.: Belknap Press, 1977).

Child, F, 'New Technology and the Labour Process', in David Knights, Hugh Willmott and David Collinson (eds), *Job Redesign*, pp. 107–41, (Brookfield, Vt.: Gower Publishing, 1985).

Chinoy, Eli, *Automobile Workers and the American Dream* (New York: Doubleday Publishing Co., 1955).

Clawson, Dan, *Bureaucracy and the Labor Process* (New York: Monthly Review Press, 1980).

Cockburn, Cynthia, *Brothers: Male Dominance and Technological Change* (London: Pluto Press, 1983).

Cooley, Mike, *Architect or Bee?: The Human Technology Relationship* (Boston: South End Press, 1980).

Dickson, John, 'Participation as a Means of Organizational Control', *Journal of Management Studies* 18 (1981) 2: 159–76.

Dreyfack, Raymond, *Making It In Management – The Japanese Way* (Rockville Center, N.Y.: Farnsworth, 1982).

Drucker, Peter, *Technology, Management and Society* (New York: Harper and Row, 1970).

Edwards, Richard, *Contested Terrain: Transformation of the Workplace in the 20th Century* (New York: Basic Books, 1979).

Ehrenreich, Barbara, and Deirdre English, *For Her Own Good: 150 Years of the Experts' Advice to Women* (New York: Anchor Books, 1979).

Fischer, Frank, 'Ideology and Organizational Theory', in Frank Fischer and Carmen Sirianni (eds), *Critical Studies in Organization and Bureaucracy*, pp. 172–90 (Philadelphia: Temple University Press, 1984).

Fraser, Steve, 'Industrial Democracy in the 1980's', *Socialist Review* 72 (November–December 1983): 99–122.

General Accounting Office, 'Advances in Automation Prompt Concern Over Increased U.S. Unemployment', 25 May 1982.

Glaberson, William, 'Is OSHA Falling Down on the Job', *The New York Times* Section 3, p. 1 (1987).

Goldman, Paul, and Donald Van Houten, 'Uncertainty, Conflict and Labor Relations in the Modern Firm 1: Productivity and Capitalism's "Human Face"', *Economic and Industrial Democracy* 1 (1 February 1980): 63–98.

—— 'Managerial Strategies and the Worker: A Marxist Analysis of Bureaucracy', in J. Kenneth Benson (ed.), *Organizational Analysis: Critique and Innovation*, pp. 85–109 (Beverly Hills: Sage Publications, 1977).

Greenbaum, Joan, *In the Name of Efficiency: Management Theory and Shopfloor Practice in Data-Processing Work* (Philadelphia: Temple University Press, 1979).

Haddad, Carol, 'Technological Change and Reindustrialization: Implications for Organized Labor', in Donald Kennedy (ed.), *Labor and Reindustrialization: Workers and Corporate Change*. pp. 137–66 (Department of Labor Studies: Pennsylvania State University, 1984).

Harris, John Howell, *The Right to Manage: Industrial Relations Policies of American Business in the 1940s* (Madison, Wisconsin: University of Wisconsin Press, 1982).

Hecksher, Charles, 'Worker Participation and Management Control', *Journal of Social Reconstruction* (January/March 1980): 55–83.

Heydebrand, Wolf V, 'Technocratic Corporatism: Toward a Theory of Occupational and Organizational Transformation'. Unpublished Paper, New York University

—— 'Organizational Contradictions in Public Bureaucracies: Toward a Marxian Theory of Organizations', in J. Kenneth Benson (ed.), *Organizational Analysis: Critique and Innovation*, pp. 85–109 (Beverly Hills: Sage Publications, 1977).

Hirschhorn, Larry, *Beyond Mechanization* (Cambridge, Mass.: MIT Press, 1983).

Kanter, Rosabeth Moss, *The Change Masters* (New York: Simon and Schuster, 1983).

Katzell, Richard, *et al.*, *A Guide to Worker Productivity Experiments in the U.S. 1971–1975* (New York: New York University Press, 1977).

Kelly, John, 'Management's Redesign of Work: Labour Process, Labour Markets and Product Markets', in D. Knights, *et al.* (eds), *Job Redesign*. pp. 30–51. (Brookfield, Vt.: Gower Publishing Co., 1985).

Kochan, Thomas, Harry C. Katz, and Robert B. McKersie, *The Transformation of American Industrial Relations* (New York: Basic Books, 1986).

Kusterer, Ken. C, *Know-How on the Job: The Important Working Knowledge of 'Unskilled' Workers* (Boulder, Co.: Westview Press, 1978).

Lipset, Seymour M., Martin Trow, and James Coleman, *Union Democracy* (New York: Anchor Books, 1962).

Main, Jeremy, 'The Trouble with Managing Japanese Style', *Fortune* (21 April 1984): 46–53.

Manwaring, Tony, and Stephen Wood, 'The Ghost in the Machine', *Socialist Review* 74 (March–April 1984): 55–83.

Marchington, Mick, *Responses to Participation at Work* (Westmead, England: Gower Publishing Co., 1980).

Marx, Karl, *Capital*, Volume I (New York: International Publishers, 1975).

McGregor, Douglas, *The Human Side of Enterprise* (New York: McGraw-Hill, 1960).

Meyer, Stephen, *The Five Dollar Day* (Albany: SUNY Press, 1981).

National Industrial Conference Board, 'Practical Experience with Profit Sharing in Industrial Establishments'. Report #29 (Boston, Mass., June 1920).

Nelson, Daniel, *Managers and Workers: Origins of the New Factory System in the U.S. 1880–1920* (Madison, Wisc.: University of Wisconsin Press, 1975).

New York Times, 'Unions are Shifting Gears but not Goals', 31 March 1987.

Noble, David. F, *Forces of Production: A Social History of Industrial Automation* (New York: Oxford University Press, 1986).

Nord, Walter R, 'The Failure of Current Applied Behavioral Science: A Marxian Perspective', *Journal of Applied Behavioral Science* 10: 4 (October–December 1974): 557–78.

Ouchi, William, *Theory Z: How American Business Can Meet the Japanese Challenge* (Reading, Mass.: Addison-Wesley, 1981).

Parker, Mike, *Inside the Circle: A Union Guide to QWL* (Boston: South End Press, 1985).

Ramsey, Harvie, 'Phantom Participation: Patterns of Power and Conflict'. *Industrial Relations Journal* 11 (1980) 3: 46–59.

—— 'Cycles of Control: Worker Participation in Sociological and Historical Perspective', *Sociology* 11 (September 1975) 3: 481–506.

Report of a Special Task Force to the Secretary of Health, Education and Welfare, *Work in America* (Cambridge, Mass.: MIT Press, 1978).

Roberts, Arthur D. and Prentice, Richard C, *Programming for Numerical*

Control Machines, 2nd edn (New York: McGraw-Hill, 1978).

Sabel, Charles, *Work and Politics: The Division of Labor in Industry* (Cambridge, Mass.: Cambridge University Press, 1982).

Salzman, Harold, 'Computer Technology and the Automation of Skill'. Paper presented at the Annual Meetings of the Eastern Sociological Society. New York, N.Y., March 1986.

Sanders, Donald H, *Computers and Management* (New York: McGraw-Hill, 1970).

Selltiz, Claire, Laurence S. Wrightman, and Stuart, W. Cook, *Research Methods in Social Relations* 3rd edn (New York: Holt, Rinehart and Winston, 1976).

—— Ibid. 2nd edn, 1959.

Shaiken, Harley, *Work Transformed: Automation and Labor in the Computer Age* (New York: Holt, Rinehart and Winston, 1984).

Siegel, Irving, and Edgar Weinberg, *Labor-Management Cooperation: The American Experience* (Kalamazoo, Mich.: W.E. Upgjohn Institute for Employment Research, 1982).

Simmons, John, and William Mares, *Working Together* (New York: Knopf, 1983).

Simon, Herbert, *The New Science of Management Decisions* (revised edition; Englewood Cliffs, N.J.: Prentice Hall, 1977).

Stein, Barry, and Rosabeth Moss Kanter, 'Building the Parallel Organization: Creating Mechanisms for Permanent QWL', *Journal of Applied Behavioral Science* 16 (1980) 3: 371–88.

Stone, Katherine, 'Origins of Job Structures in the Steel Industry', *Review of Radical Political Economics* (Summer 1974): 113–73.

Taylor, Frederick Winslow, *Scientific Management* (Westport, Ct.: Greenwood Press, 1972).

Thomas, Robert J, 'Participation and Control: New Trends in Labor Relations in the Auto Industry'. Center for Research on Social Organization. Ann Arbor: University of Michigan Working Paper #315.

Thompson, Paul, *The Nature of Work: An Introduction to Debates on the Labour Process* (London: Macmillan Press, 1983).

—— and Bannon, Eddie, *Working the System: The Shopfloor and New Technology* (London: Pluto Press, 1985).

Tulin, Roger, *A Machinist's Semi-Automated Life* (San Pedro: Singlejack Books, 1984).

U.S. Congress. Office of Technology Assessment, *Manufacturing Automation: Employment, Education and the Workplace* (Washington, D.C. OTA-CIT–235. April 1984).

U.S. Department of Labor. Bureau of Labor Statistics, *Occupational Projections and Training Data: A Statistical and Research Supplement to the 1986–87 Occupational Outlook Handbook*. April 1986. Bulletin 2251.

—— Labor-Management Services Administration, 'Starting Labor-Management Programs'. Superintendent of Documents (Washington, D.C.: U.S. Government Printing Office. September 1982).

Vaughan, Edward, 'Industrial Democracy: Consensus and Confusion', *Industrial Relations Journal* 11 (1 March/April 1980): 50–6.

Walker, Charles, and Robert Guest, *Man on the Assembly Line* (Cambridge, Mass.: Harvard University Press, 1952).

Wells, Donald M, *Empty Promises: QWL Programs and the Labor Movement* (New York: Monthly Review Press, 1987).

Wilkinson, Barry, *Shopfloor Politics of New Technology* (London: Heinemann Books, 1983).

Windolf, Paul, 'Industrial Robots in the West German Automobile Industry', *Politics and Society* 14 (1985) 4: 399–495.

Witte, John F, *Democracy, Authority and Alienation in Work: Workers' Participation in an American Corporation* (Chicago: University of Chicago Press, 1980).

Wrenn, Robert, 'Management and Work Humanization', *The Insurgent Sociologist* 11 (Fall 1982) 3: 23–38.

Yankelovich, Daniel, and John Immerwahr, 'Putting the Work Ethic to Work', *Society* (January/February 1984): 58–76.

Zimbalist, Andrew, ed, *Case Studies on the Labor Process* (New York: Monthly Review Press, 1979).

Zuboff, Shoshana, 'New Worlds of Computer-Mediated Work', *Harvard Business Review* (September/October 1982): 142–52.

Index